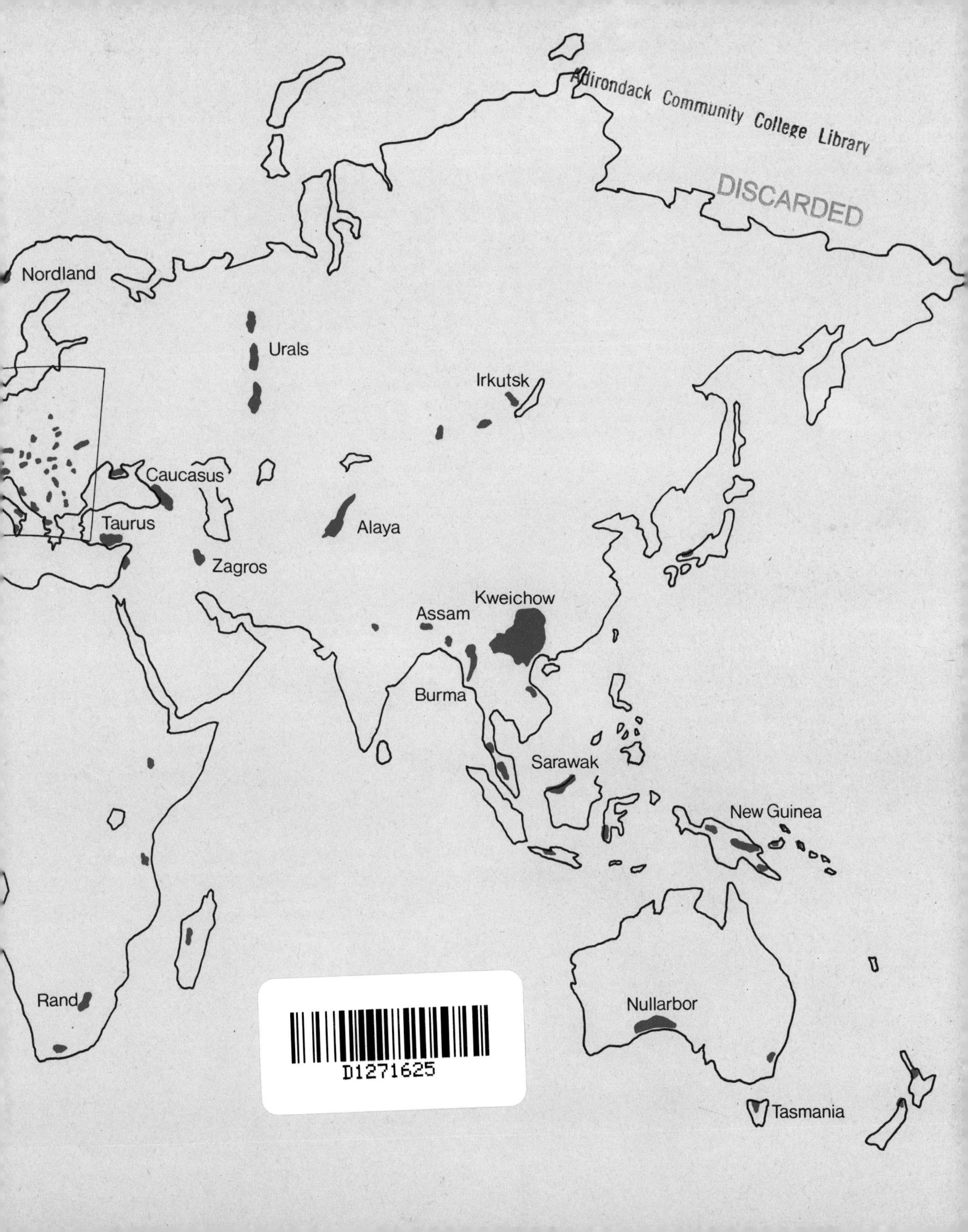
Nordland
Urals
Irkutsk
Caucasus
Taurus
Alaya
Zagros
Kweichow
Assam
Burma
Sarawak
New Guinea
Rand
Nullarbor
Tasmania

CAVES

CAVES

Tony Waltham

Crown Publishers, Inc.

Author's acknowledgements
This book would never have been possible without the help of the many caving companions who accompanied the author underground. Grateful thanks are extended to them all. In addition the author would like to thank Dave Checkley who read and criticised the draft of the manuscript on life in caves, Alan King who did the same with the section concerning archaeology, Penny Brown who helped with the picture research, and Bill Renshaw who patiently acted as model for many of the photographs.

Frontispiece: Kalvarija Chamber in Jugoslavia's Krizna Jama

Front end-paper: Cave areas of the World

Back end-paper: Cave areas of Europe

Library of Congress Catalog Card Number 74–79884

Designed by Paul Watkins

Published in the United States of America 1974
by Crown Publishers, Inc., New York
and in Great Britain 1974
by Macmillan London Limited
London and Basingstoke
Associated companies in New York, Toronto, Dublin, Melbourne, Johannesburg and Delhi

Printed in Great Britain by Jolly and Barber Ltd., Rugby
and by BAS Printers Ltd, Over Wallop, Hampshire

Contents

1 The World of Caves

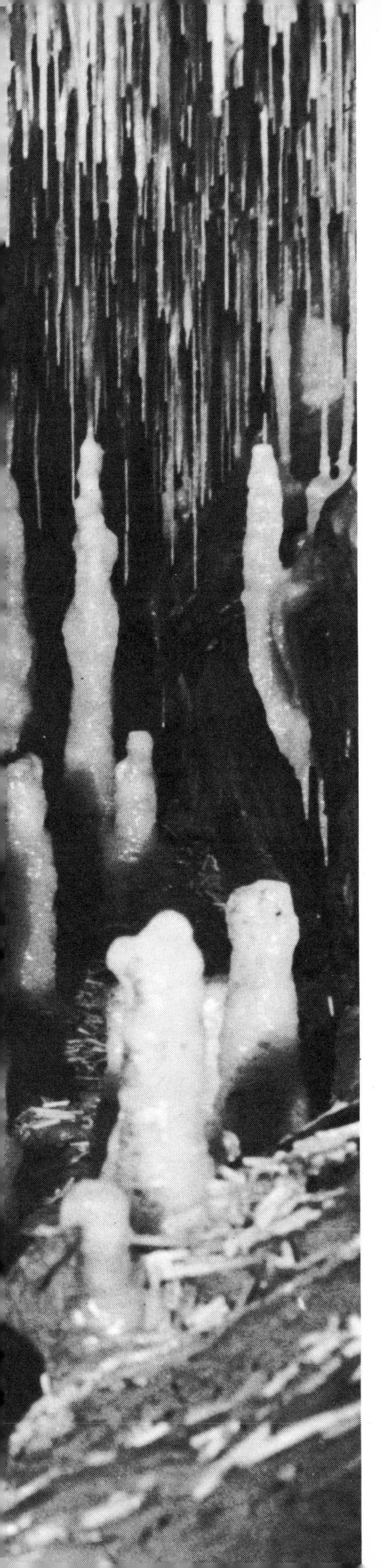

The stream crashes noisily into the clean-washed chamber. A wire ladder is unrolled and lowered, and the cavers climb down in the spray. The chamber is nearly three hundred feet below the windswept Pennine moors, which the cavers left behind two hours before, to follow the stream down the narrow galleries and deep shafts of Strans Gill Pot – just one of England's many cave systems. In the far wall of this cavern the stream runs away into a narrow rift and plunges to further depths. But the beams of the cap-lamps cut into the darkness, and show the cavers their way on. A ten-foot climb up the right-hand wall there is a passage at roof level. It is low and jagged, but dry; the noise of the waterfall is left behind as they crawl onwards. Less than a hundred feet further on, the floor drops away and the ceiling rises into a long chamber. A channel is cut in the hard dry mud floor, and the arched roof is adorned with thousands of slender straw stalactites, each a yard or more long. They are pure white, and sparkle in the beams of the lamps. This silent and beautiful cavern provides a vivid contrast with the stark rock of the noisy stream passages that lead to it.

Contrast and variety – these are the very essence of caves. Caves are frequently described as networks of holes in the ground. This is true of course, but it is a rather callous way to speak of a unique and fascinating world. There is nothing on the surface of the Earth which is at all similar to this world below. Caves consist of miles of passages, great caverns hundreds of feet high and wide, with roaring stream passages and fabulously decorated stalactite and stalagmite chambers. There are even some rather nasty, muddy little tunnels – to fit the popular misguided concept of a cave. And until a caver wanders in with a lamp, this whole peculiar world is in total darkness.

To many people, darkness is disturbing. And yet few can have experienced the absolute and total darkness of a cave. Beyond the entrance, there are no little chinks of light, no reflections, no glows from distant houses – there is no light at all. A hand held an inch in front of the face is totally invisible. The only lights are those that the caver takes in for himself. The helmet-mounted lamp moves around with the eyes and gives the false impression that the cave is lit all over, and not just in a tiny wedge in the beam of the lamp. Even a properly equipped caver will sometimes, perhaps while he is waiting alone for his colleagues, put out his lamp to experience the total darkness of caves. And it is a very odd experience: with no light at all, he has only his memory to tell him what surrounds him, he will probably soon feel cold, he may lose his sense of balance, and in a short time he may begin to feel uneasy. Only the really experienced caver can feel relaxed in this situation; most would shortly, almost involuntarily, go to turn on their light again.

Exploring caves

Dowkerbottom Cave (far left) in the English Pennines invites the explorer into the darkness, while a long ladder climb takes the caver down a finely fluted shaft in Ghar Parau (left) in the Zagros Mountains in Iran. In the Spanish cave of Calducuin (below) the only way along the stream passage is by swimming in the cold water. Strans Gill Pot in the English Pennines has a series of really tight squeezes (right) in the entrances, but leads down to the beautifully decorated Passage of Time (below right)

And then there is the silence. Many caves are completely silent – an emotionless feature which matches the darkness. In others the sound of dripping water, or the dull rumble of a distant stream, or even the tremendous roar of an adjacent waterfall may be heard. But these are all sounds which are constant, unless there is a rainstorm on the surface above, and after a time they seem totally uniform and featureless to the listener. It is a sobering thought that this almost dimensionless, dark environment has probably remained unchanged in most caves for five thousand years or more.

Everyone is surprised when they enter a cave for the first time. Some may turn back in sheer horror, a few may experience claustrophobia, but most will continue in incredulous amazement. They will probably be surprised to find that not all caves are 'tight muddy little tubes in the rock', but that many consist of lofty galleries, with the roofs lost in the darkness, and containing

wide, swift, underground rivers. Clean rock walls and sheets of stalactites are usually more abundant than mud. Most people who first enter a cave do so in a show cave – one that is cleverly illuminated by hidden electric lamps, and provided with a convenient footpath. But this is not the true cave environment. The darkness, the mystery, the strangeness of a world in which man is an intruder – all these aspects of an untouched cave are lost in a show cave. Caves are a unique feature of this planet; they contain much to fascinate man, yet he can barely appreciate them. After all, who can really visualize a hole in the ground hundreds of feet below the surface? Even the caver who looks up at the roof of his passage finds it difficult to understand that there are tons of rock above his head.

The best way to visualize the depth of a cave is to stand on the surface, in daylight, above a cave. Look at how small people appear who are, say, two hundred yards away; and then look

Stalactites and stalagmites decorate a small passage in the Père Nöel Cave in Belgium

down at the ground and try to imagine people appearing that small as they are walking around in dark caverns far below the ground!

Caves may seem timeless, but in fact change does take place. The streams which flow through so many of them may be quite commonly the very ones which originally formed the caves. On the other hand, the cave may have been formed by some ancient and long-gone underground river and a different stream may have come to flow through it much later, for the waterways in limestone are frequently changing their paths. But, whatever its course, the water is all the time steadily dissolving away the limestone, and carrying away the rock as invisible solutions – perhaps, in some areas, to be redeposited as the 'fur' in someone's kettle. The amount of rock so moved from the walls of a stream cave may only be a hair's thickness each month – but over ten thousand years this can form a spectacular cave. So the caves do change over time – admittedly very slowly – and indeed this is how they are formed. Visualize a river, on the ground, with its myriads of tributaries. Then think of this underground, so that it forms not a two-dimensional maze, but a three-dimensional one. Then imagine a roof over each stream channel, remembering that where streams change their course the abandoned channels are not washed away or obliterated by rain or wind. And that network of old and new channels is a cave system in limestone.

To many, this water worn sculpture of the rock is the real attraction of caves. The deep meandering canyon caves, the massive chambers and the intricate spongeworks of tube caves each have their special characteristics. A geologist can spend many a year determining the long sequence of events that must have been responsible for the origin of the world's great cave systems. Why is this chamber so large? Why does that passage suddenly spiral underneath itself? Why does that stream start in that valley and flow through to this one? The questions are endless, and the unanswered ones only increase the mystery of caves. Each cave has its own story to tell, and its own secrets to hide.

But to most people it is what happens in the cave after it has been formed that provides the main fascination. Light, essentially sunlight, is obviously important to the patterns of life and the processes which take place on the surface of the Earth. But just how important is dramatically demonstrated in caves, for the total lack of light is directly responsible for caves' two most characteristic features – the stalactites and stalagmites, and the unique animal life.

Rainwater that percolates through the tiny fissures in limestone will frequently become laden with dissolved calcite (calcite is the mineral that forms both limestone and stalactites), and this percolating water will eventually reach a state where it is saturated – that is, it can absorb no more of the mineral. Then if ever that saturated water drips or seeps out of the rock, and if conditions of temperature and the chemical composition of the air are correct, it will deposit some of the dissolved calcite – as a stalactite or a stalagmite. Clearly, a hole in the rock, a cave, provides an ideal site for this deposition by dripping water, but it can also take place on cliff faces, for example, out in the open air. The stalactites inside a cave are commonly beautifully clean, pure white or sometimes colourless, and maybe sparkling in the light of a lamp. Yet calcite deposits out in daylight, besides being much rarer, are nearly always dull and dirty, and this is largely due to algae growing in the films of water on their surfaces. But algae will not grow in darkness, so it is the lack of light in caves which is normally responsible for the sparkling whiteness of their calcite decorations. The problem that then arises (for the caver, or the chemist, or the inquisitive visitor) is just how the great variety of stalactite and stalagmite shapes developed in the world's caves.

Dripping water from joints in the bedrock have formed the straight lines of stalactites and stalagmites in the Père Nöel Cave, Belgium

A party of tourists sail by boat through the Waitomo Cave in New Zealand to see the unique colonies of glow-worms hanging from the cave roof

It is appropriate that one of the most abundant inhabitants of caves – the bat – should be one of the extremes of the animal world. These usually small, flying mammals feed at night, and need a safe place for their daytime sleeping periods. Caves provide an ideally safe daytime dormitory; bats have been able to make use of them by virtue of an unusual adaptation to a lightless environment – they have a unique built-in radar device which enables them to fly around in total darkness. Like bats, almost all animals that live in caves have a characteristic feature related to the darkness of the underground, and the bat is not even the most peculiar of the cave dwelling animals. There is a blind shrimp, which is so colourless that all its internal organs are clearly visible as it swims past – it may sound like a piece of science fiction, but it is a common inhabitant of caves almost all round the world. Indeed cave animals are a unique group, generating a great deal of research among the world's biologists.

Man himself is completely opposite to the cave-adapted shrimp, for he is dependent on light. So to him the cave is a hostile environment. But man also has the ability to make and build; he can make light, his own artificial sunlight. Consequently he can enter a cave, and, albeit temporarily and rather precariously, transform it into an acceptable environment. But for tens of thousands of years, man's own home-made lights were rather crude affairs – dull, smoky, unreliable brushwood torches and, later, simple tallow torches. So still it was only the venturous few who dared to roam deep into the dark zone of a cave. Yet the caves provided ready-made shelter for primitive man. So he did go in them, indeed he lived in them, but only in the entrance zones. And as the cavemen wandered from one cave dwelling to another, they left behind odd materials or simple implements – left them behind in the almost changeless cave environment. Consequently these remains of early man are better preserved than anywhere on the surface of the Earth, and provide a massive data bank for study by archaeologists. But primitive man's real gems were left behind by the venturous few who explored further into the caves. There, in peaceful silent chambers, they painted on the walls and left an unparalleled record of their life styles. Even today these painted caves are still being discovered and laid bare to human eyes for the first time in perhaps thousands of years.

Men always had this spirit of adventure, and there has been a natural development of interest beyond his early instinct just to wander into the recesses and dark zones of his early cave-homes. This interest has grown rapidly within the last century, mainly for two reasons: the greater amount of leisure time available for modern man, and the inventions which have slightly tamed the hostility of caves, the electric light for instance.

The Big Room in Carlsbad Cavern, USA (above) is decorated by stalactites and stalagmites which match its size. The Grand Gallery in the Berger Cave (right) is aptly named – crawling and squeezing are just not involved in a descent of this magnificent French cave

Consequently caving has become a sport. It combines the excitement of danger with the attraction of seeing the unique underground world, and it is different. The Everests of the mountaineering world are there to be seen – they provide a challenge and men climb them 'because they are there'. The underground world provides a similar challenge and fascination, though not quite on the same scale. Yet the Everests, even the Snowdons of the underground, provide an extra complication – they cannot be seen, they are unknown until they are visited. The unknown always fascinates man, and this is one of the big attractions of the sport of caving. Even in Britain, at the present time, a caver can discover and explore a new cave; he can 'tread ground never before trodden by man'.

It is a tremendous excitement for a caver to explore for the first time even a hundred feet or so of quite ordinary cave passage. But occasionally a really magnificent cave is found – and then it is more than just a fine discovery. One such is the Carlsbad Cavern. It lies in the semi-deserts of the southwestern United States, and its entrance is high in the barren Guadalupe Mountains. A huge passage leads down to the fabled Big Room. Averaging three hundred feet high and six hundred feet wide, it is over a thousand yards long. Massive stalactites cascade from the walls and ceiling, and the rocky floor is broken by mountainous stalagmites. Carlsbad is not only a wonder in the world of caves; it is one of the wonders of the world.

Caves are different

Part of the attraction of the underground is that no two caves are alike. Deep inside the great Welsh cave, Dan-yr-Ogof (left) the stream flows quietly across a gravel floor. But in N.W. England the lower streamway of Lancaster Hole (right) foams over a series of noisy cascades. And in Pippikin Hole (below), also in the English Pennines, there is no stream to break the silence of these stalagmite-strewn caverns

The calcite deposits have spectacularly decorated this part of the Gournier Cave in southern France (far left). Stalactites hang from the roof, and flowstone dams hold back a series of clear green pools along the passage floor. A single drip from the passage roof in G.B. Cave in the English Mendips (left) has formed a cluster of stalactites clinging to a ledge on the wall

2 The World of Karst

Caves have a certain rarity value because they do not occur in all parts of the world. Instead, they are nearly all found in one distinct type of limestone terrain, known as 'karst'. Even without its caves, karst is a unique and remarkable landscape.

The word 'karst' is a distortion of the Slovenian word *kras*; the *kras* is the name of a region which now straddles the Jugoslavia–Italy border just east of Trieste. It is an area devoid of surface rivers, a land of bare white limestone surfaces with only a thin patchy soil, of peculiarly fretted rock and gaping cave entrances – in fact it is one of the finest karst regions in the world, and has given its name to all of them.

Karst is now defined as a landscape which has been developed by water solution of the rocks. And therein lies the importance of limestone in karst landscapes – it is by far the most soluble of the commonly occurring rocks. The Earth's surface form as we know it has been carved out of the solid rock by long periods of natural breakdown and removal – known jointly as erosion. The three main erosive agents are water, wind and ice, and, of the three, water is by far the most important. Yet water can only erode and wear away rocks effectively when it is charged with rock fragments. The real erosive power of both mountain stream and lowland river lies in the sediment being carried along by the water – acting like a great rasp on the river bed.

Consider now any mass of rock. Water erosion takes place on its surface, and yet nearly all rocks contain networks of joints, or fractures, and these too, except in a desert, will be full of moving water. However, in a fracture only a few microns wide, surface tension slows the water down so much that it is certainly incapable of carrying along any sediment to erode and enlarge the joint; so the opening stays very, very narrow. But in limestone the situation is different, for there the water does not need to wear away the rock. Instead it can just dissolve the rock, and carry it away in solution, regardless of its speed. And so the fractures are enlarged, and as they get wider their water moves faster, thereby eroding more efficiently. In time, the fractures are opened up to an enterable size – to become known as caves.

Eventually this network of caves reaches a size when it can take all the natural drainage, and any surface rivers therefore disappear. Consequently the caves are not only an integral, but an essential, part of the karst landscape. Any landscape is dominated by its drainage systems, and caves influence the drainage pattern in the most extreme way possible – they entirely remove it from the surface and take it underground.

There are other rocks, besides limestone, which are soluble in water. Gypsum and rock salt are the two most important, and they can contain both karst and caves. However they are both very uncommon in comparison with the worldwide distribution of limestone. Whereas nearly every country has some limestone karst, few have any extensive areas of gypsum or salt. Russia is perhaps the most marked exception, with long cave systems known in both limestone and gypsum, and great karst regions formed in limestone, gypsum and salt.

Caves also form in some completely insoluble rocks, but the landscapes of such regions cannot be called karst. Such a rock is basalt – the most common type of rock solidified from a lava straight out of a volcano. A single lava flow is a stream of molten rock pouring from a volcanic vent. It solidifies into basalt merely by cooling down, and naturally the top, exposed parts of any flow cool down first. Consequently a solid crust may grow out steadily from the edges to cover over the lava stream, which in its inner part is still molten and flowing. The molten core may then flow away and leave a lava tube inside the basalt. Some such lava caves are known to be many miles long, in volcanic areas in Iceland, the Canary Islands and the United States. Other types of caves in non-karstic landscapes include sandstone shelters and sea caves. The former are carved out of sandstone cliffs by wind action – effectively sand blasting – in a desert environment, and the latter are merely hammered out of sea cliffs by the pounding effect of wave action. Both these types are

Lava caves

New Cave in the Cascade Mountains of Washington, USA (left) is a typical lava tube with a rounded roof and a rough lava form. The Surtshellir Cave (above) in the Iceland lavas is entered where the thin roof has collapsed into the tube. A cave in Kenya's Mt Suswa (right) is adorned with lavacicles, where some of the molten rock dripped from the roof after the main mass of lava had drained out of the tube

normally only shallow rock-shelter caves; but, if the environment of the sandstone caves is not too harsh, or if sea caves are left abandoned on land by a fall in sea-level, both can provide convenient shelter for primitive man and so occupy the archaeologist of today.

The single feature which most distinguishes true karst is the lack of surface drainage. The presence of the limestone caves – nature's most efficient built-in drainage system – means that rainfall disappears into the ground, either just where it lands or after collecting into very short disconnected stream courses. In addition, the almost complete dissolution of limestone in rainwater means that soils are either very thin or completely non-existent in karst regions because there is little insoluble residue left on the surface to form the basis of soil. Consequently, outside the tropical regions, karst tends to be either thinly vegetated or left as bare rocky ground. This, and the lack of surface drainage, explain why a number of the French karsts are known as deserts, and even in the tropically wet karst of Jamaica, cattle can sometimes be heard moaning because of their thirst – their water supply is separated from them by three hundred feet of rock.

In detail karst landscapes may vary quite considerably. The Causses of southern France represent one extreme – rolling, relatively featureless country with no rivers and few trees. The

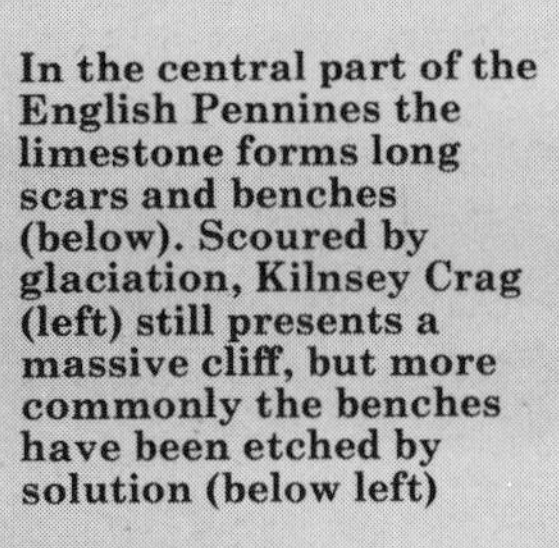

In the central part of the English Pennines the limestone forms long scars and benches (below). Scoured by glaciation, Kilnsey Crag (left) still presents a massive cliff, but more commonly the benches have been etched by solution (below left)

surface is an endless sea of broken limestone rubble with little soil and only thin short grass growing on it. Winter frosts and summer heat have broken the limestone up, but the percolating rainwater cannot carry it away fast enough purely in solution – so it remains on the surface. In contrast, exactly the same rock forms great cliffs and smooth rock surfaces in the karst areas of the French Alps – uninterrupted scars of limestone run right round the mountains and the bare rock benches are practically devoid of all vegetation. The great contrast is due to glaciation. In the French Alps, and similarly in such regions as the Austrian Alps and the English Pennines of Yorkshire, glaciers swept all the broken rubble away during the last Ice Age. The solid scoured surfaces left behind by the glaciers, less than ten thousand years ago, have not broken up again due to weathering. Instead only solutional etching and carving have modified these benches to give the well-known limestone pavements that can be seen today – great slabs of limestone separated by solutionally opened joints, or 'grykes', which absorb all the drainage. Both modern and past climates have considerable effect on the forms of development in limestone karsts.

There is normally little or no running water on the surface of karst landscapes; and yet valleys, admittedly dry, are common in limestone areas. How then are they formed? Some have been excavated by glaciers – many of the Yorkshire Dales in the English Pennines are of this origin; but glaciated valleys are easily distinguished and most karst valleys are not glacial. A popular, old, explanation for many of the steeper-sided valleys is that they are collapsed caves. Cheddar Gorge in the English Mendips is frequently cited as an example. But this is not true – and the evidence lies in the shape. Firstly, Cheddar Gorge is far larger than any caves known in the area, and it also has a gently curving, uniformly branching plan – completely unlike the geologically controlled networks of the caves. Its long profile is also unlike that of a collapsed cavern, in that its floor has a continuous gentle slope whereas a cave system would descend steeply at its head and be nearly horizontal lower down – as indeed are the cave systems opening out of the bottom of the gorge.

Instead, the dry valleys of limestone karst areas have nearly all been cut by surface rivers – they do after all have the characteristic size and shape of normal river valleys, except that they have no rivers. So they are fossil features – cut by rivers which have now abounded them and gone into cave systems. But why did the rivers ever flow on the limestone surface? The answer to that question lies in the limestone terrains found on some of the arctic islands of northern Canada. There great rivers cascade over bare surfaces of pure limestone; they cannot flow underground, because the ground is frozen, and all the fissures and cave segments are blocked solid by ice. This kind of environment is described as 'periglacial' – affected by, but not covered by, glaciers – and the ground is in a state of 'permafrost'. It is permanently frozen, except that during each summer the top few metres melt when the surface rivers carry away the water from the winter snow, and carve their valleys a little bit deeper. During the Ice Ages of the past million years nearly all of Europe and North America were for some periods of time in the permafrost zone, and it is from these times that the dry karst valleys originate. The English Pennines, the American Appalachians and the French Alps and Causses all had their dry valleys formed in this way. Cheddar Gorge is just a rather rocky valley of this type. The coombes in the Chalk Downs of S. E. England are similarly periglacial in origin, and now dry.

Collapsed caverns are indeed both quite rare and relatively small features. Very few of undisputed collapse origin are known to exist and most of those are in the sides of steep, shallow valleys, where water has easily cut down under scars to form short, large-diameter caves, with their roofs ready to collapse. Furthermore many of these have cave openings at their backs, so that they are not really true 'collapsed-cavern valleys' but are merely collapsed cave entrances.

Karst regions also contain some of the world's finest gorges. The 'Grand Canyon of France' – the Verdon Gorge – and the 'Grand Canyon of Greece' – the Vicos Gorge – are both cut into limestone. The River Vicos, in the extreme northwest of Greece, flows through its magnificent nine-mile gorge between almost unbroken lines of cliffs nearly three thousand feet high. Much of the excavation of the gorge probably took place during the

The world's greatest karst region in Southern China contains vast areas of deeply fluted limestone – some of the most spectacular is in the 'Stone Forest' in the Yunnan province

Limestone cliffs loom over the dry riverbed of the Vicos Gorge in Greece

periglacial periods of the last million years. Why then have the cliffs not been worn back since that time, to give a normal V-shaped profile to the valley? A gorge of that age in any other rock than limestone would no longer have such well-preserved cliff-lines. Normally this opening up of the gorge would be the natural result of rainwash and small tributary stream erosion. But limestone is different – it 'devours its own agents of erosion'. The rainwater and small streams all flow underground and leave the surface unmarked. This even happened to some extent during the meltwater cutting of the gorge, when rainfall water did not flow over the gorge edge but short-cut down through the shallow seasonally melted zone in the limestone. Today the Vicos Gorge is dry in the summer because the river sinks at the top end of the gorge, where it first meets the limestone, and emerges at a great spring near the lower end of both the limestone and the gorge. This flooded unexplored cave system beneath the gorge is as yet very immature, for it is incapable of taking the high winter flows of the river – the gorge temporarily becomes an active riverbed again for a short period each year.

Rather analogous to the Vicos Gorge in winter, but on a much smaller scale, is the valley of the River Dove through the Peak District limestone in central England; for it flows as a surface river in a periglacial valley through solid cavernous limestone. It doesn't sink and flow underground across the limestone outcrop. This is because its headwaters collect on an adjacent area of sandstones and clays and it is a sizeable river by the time it meets the limestone. It is then so large that it cannot just slip down a joint in the rock, and due to its low overall gradient across the limestone outcrop it does not have the power to open up new caves in dissolving away an underground route. Water is basically lazy, and it will always take the easiest way down its course – in this case over the surface instead of via a tight devious network of fissures in the limestone. A small fraction of the Dove's water does leak underground to reappear lower down as springs; over time this will steadily enlarge a cave and so take an increasingly large share of the river. The Vicos River is half way through this process – its cave route takes all its summer flow but is not large enough for its winter flow – while other valleys have now completely dried up in this manner.

This situation, where a large surface river has to flow across limestones with very little overall gradient, explains the existence of most of the rivers in karst regions. The Tarn and Verdon rivers in southern France are both of this type. And the eventual diversion of such rivers to underground routes provides many of the dry valleys found in regions which have never been subjected to periglacial erosion. It is however, noticeable that the tropical karst areas, such as in Jamaica, contain many fewer dry valleys.

A closed depression – a hollow with no downhill outlet – can exist in almost any landscape, but in most cases it will fill with water, and is then known as a lake. Dry closed depressions can only occur in two special cases: where there is no water, as in a desert, or where the water can leak out through a hole in the floor of the hollow, in a karst region. All the closed depressions that are so characteristic of karst landscapes are related to cave systems which provide their drainage – though in many cases the caves may not be large enough for man to enter and explore. Most karst depressions range from a few yards to a hundred yards in diameter and are roughly conical or saucer-shaped; they are then known as 'dolines'. This is the Slovenian word for a valley, but in the Jugoslavian karst there are almost no real valleys, all they have are short closed valleys – the dolines. Nearly all the world's karst regions have dolines of some size and shape and in some abundance. A large proportion of caves and potholes have their entrances in dolines; indeed, where is the boundary between very steep-sided dolines and wide-open potholes? The Linton Park Lighthole is a massive closed depression in central Jamaica – it is nearly half a mile across and is completely ringed by vertical cliffs 300–650 feet high.

Some of the most remarkable dolines are in fact full of water. In the western U.S.A. lie the Roswell Bottomless Lakes of New Mexico. They are part of a group of about twenty steep-sided

A network of large dolines breaks the surface of the North Parau Plateau in the high-altitude fossil karst of Iran's Zagros Mountains

dolines, twelve of which have lakes in them, and despite their title the deepest contains less than 130 feet of water. They are between 50 and 100 yards in diameter. They are particularly conspicuous because this part of New Mexico is virtually a desert, with only the sluggish Pecos River flowing by, over half a mile away. Although the Bottomless Lakes now act as sinkholes and absorb any rainwater that arrives, they were formed by upward leakage of groundwater from a buried water-carrying rock – their flow of water, though now very small, is the reverse of what it used to be.

Probably the nearest thing to a bottomless lake is the lake of Crveno Jezero in southern Jugoslavia. It lies in a depression 400 yards across, ringed by cliffs 500 feet high. Its depth has been measured as an incredible 880 feet! Perhaps it should not really be known as a doline but as a flooded pothole. Also of massive proportions, but now completely flooded, are the Blue Holes of Andros Island in the Bahamas. These lie in the reefs around the edge of the island; average depth of the sea there is only a few feet, but it plunges to 300 feet in the Blue Holes – almost vertical sided holes up to several hundred feet across. They too are submerged dolines, formed when the sea level was 300 feet lower than now, when the Ice Ages had locked great volumes of water in massive ice caps.

Largest of all the closed depressions in karst are 'poljes' – great flat-floored valleys completely surrounded by bare limestone mountains. Some of those in Jugoslavia are over thirty miles long. They take their name from the Serbo-Croat for a cultivated field, for in the mountainous Dinaric karst the only land fit for cultivation is on the sediment covered floors of the poljes. Their origins are still not fully understood, but are essentially related to the geological structure. Their most remarkable feature is their hydrology: typically they have large springs around their margins, producing streams and rivers which flow along the alluvial floor to disappear into 'ponors', or sinkholes.

This in itself would not be remarkable in a karst region, but the capacity of the ponors is not always large enough to absorb all the water produced from the springs. Consequently the poljes are subject to seasonal inundation. The length of the inundation period depends upon the size of the ponor drainage channels and the seasonal intensity of the rainfall, which feeds the springs; some poljes are nearly always dry, while others have almost permanent lakes in them. But the poljes which are not permanently flooded provide the best agricultural land in the karst of Jugoslavia – as long as the farmers correctly anticipate the date of the annual inundation! Popovo Polje, just inland from Dubrovnik, is typical in this respect. In winter it is a huge lake.

In summer it is a mass of intensively cultivated fields – but even then it does not look like a normal valley. The road is perched sixty-five feet up on the rocky hillside, and the villages are similarly crammed into rocky enclaves – seemingly avoiding the convenient flat land below, but in fact merely being prepared for the winter.

Last but not least of the really major components of a karst landscape are the caves themselves. Cave entrances vary from the obscure to impressive. Some karst regions have barely an entrance visible, while other areas seem completely perforated with openings of all shapes and sizes – each one leading into blackness, a perfect challenge for man's instinct for exploration.

The quarry-like opening of Hull Pot in the English Pennines is floored with stones and boulders which absorb a large river in times of flood

The title for the largest cave entrance in the world is often disputed. One that has been frequently quoted is the Bournillon Cave in the Vercous Mountains near Grenoble, France. It is a 'resurgence cave' – for a sizeable river emerges in times of flood – and it opens at the base of an immense vertical cliff of white limestone. The opening is over 260 feet high and 100 feet wide. A footpath beside the river bed leads into the cave from the valley below, and the entrance is in a corner – not easily seen from any distance away. The visitor may follow the path and not notice he has entered the cave. But he will notice that the thick undergrowth on either side of the path suddenly stops, in a remarkably straight line. Beyond is only bare rock and rubble. Looking up from that line, the end of the cave roof is vertically overhead – 260 feet up above. There is no vegetation further in because there

Dams of travertine

Calcite deposits are common enough inside caves; but in the open air though much rarer they tend to be on an impressive scale. The Mammoth Springs of Yellowstone Park, USA (above right) pour out water saturated with lime, so that they have built up these great terraces of travertine – a form of calcite. On each terrace there is a pool and the water cascades down the whole series, continuously depositing layers of calcite as it flows over each of the self-building dams. The same process has formed the barriers around the lakes of Band-i-Amir (below right) in central Afghanistan. Surrounded by arid limestone plateaus, the water emerges from springs, flows into the lakes, deposits the travertine as it flows over the dams, and then sinks into the valley floor gravels. No calcite deposits break the stark grandeur of the vast exit from the Bournillon Cave in Southern France (left). The figure (ringed) shows the scale

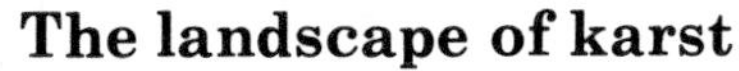

The landscape of karst

The great chasm of Bridge Pot (below) in the limestone mountains of Lebanon, formed long ago by a sinking river, is now dry – just one more hole in the karst. The massive shaft of the Fontaine de Vaucluse (left) in the Rhône valley of France never has water entering it. Instead, during the wet season, water under pressure flows up and out of it; in summer there is just a lake fifty feet down the shaft, while the winter level is marked by the white line on the cliff behind. Similar seasonal flooding affects the Kimbos Polje in Turkey (below left). In summer it is a dry valley surrounded by limestone mountains, but its underground drainage outlets cannot accommodate the winter floods – so each year it becomes a temporary lake. Typical of so many karst regions are the deeply fluted limestone blocks, here seen (below centre) on the Astraka Plateau of Greece

The ice tunnel in Canada's Athabaska Glacier is similar to those found in many of the world's glaciers. Formed originally by summer meltwater flowing down crevasses and through the heart of the glacier, it is now abandoned and dry, with its glistening ice walls carved into scallops by eddies in the wind blowing through

is no rainfall. The path continues into this gigantic passage for another few hundred yards, and there the limit of daylight is met at a narrowing of the passage on a corner. The cave continues in perpetual darkness. Even the Bournillon must now concede any size records to the more recently discovered Kara River Cave near the west coast of Bougainville Island in the East Indies. The entrance to this is very nearly 330 feet high and an incredible 360 feet wide – so vast indeed that it frequently has its own cloud systems inside it!

Nearly at the opposite end of the size scale lies the entrance to Pool Sink in northwest England. It is a horizontal tube only one and a half feet in diameter, in the bank of the periodically dry Ease Gill streambed. Behind a large rock, it could be easily hidden by a few bits of stream-transported vegetation. Yet who could believe that beyond this tiny opening lies a network of over eighteen miles of cave passages.

Completely different again are the vertical openings into caves – the pothole entrances. These, too, range in size from narrow slits in the rock, barely wide enough to squeeze down, to immense shafts. High in the forest-covered mountains of Xilitla in eastern Mexico, a now famous pothole was discovered by American cavers only in 1967. Known as the Sotano de las Golondrinas, its entrance is about 150 feet in diameter – and 1200 feet deep! Furthermore it opens out at depth to a floor area of 980 by 490 feet – making the caver's descent down the centre of this vast chamber even more spectacular.

To add to their splendour, many potholes have streams or rivers cascading into them. In the dry winter season, the Harpan River Cave in Nepal has a fine waterfall plunging the entire 150 feet depth of its upper entrance. It is an impressive sight in the chamber at the bottom of the shaft, where the cascade is still lit up by daylight. But the whole scene is even more remarkable at the height of the monsoon. For then the massive flow of the river completely fills the 35 foot diameter shaft, which acts merely as a great plug-hole in the riverbed; the overflow drops down an adjacent section of the shaft.

More than twice as deep as the Nepalese pothole is one at Vaucluse in southern France. But here the water does not go down a great cascade; instead it flows up, under pressure. It is the Fountaine de Vaucluse, source of the River Sorgue. The water wells up this great shaft from the unknown depths of the limestone mountain, coming to the daylight again after an underground journey of up to twenty five miles from its distant sinkholes. Vaucluse is perhaps the most remarkable of the many thousands of springs in the world's karst regions – and behind each spring must lie a cave system transmitting the water through the ground.

Cave rivers

The open black shaft of Rumbling Hole (right) in the English Pennines no longer carries a stream into the cave below. But the entire flow of the Harpan River is engulfed by a deep shaft (far right) in the foothills of the Nepal Himalaya. The waterfall makes the descent of the Harpan River sinkhole a spectacular challenge (below right). In south Wales the river flows smoothly into the wide entrance of Porth yr Ogof (below)

Climatic effects on karst are not restricted to the glacial scouring of limestone pavements in mountainous regions. Weird and exotic landscapes occur in the tropical karst areas, where the high rainfalls and temperatures lead to extreme types of solutional landforms. In Jamaica the forest-covered limestones are carved into an endless mass of cockpits – rather like dolines – and conical hills. This is frequently called 'egg-box topography' – a most apt comparison for a chaotic landscape with virtually no flat land within it. A variety of tropical karst even more bizarre than the cockpit karst is the 'arête' karst of New Guinea. This consists again of a dense mass of closed depressions, each hundreds of yards across; but here they are separated by jagged, almost razor-sharp, vertical sheets of bare limestone – the

Tropical karst is distinguished by the isolated limestone hills. In the Philippine Islands (below) these are vegetation covered rounded cones. But in the Kwanghsi karst of Southern China the limestone forms spectacular towers (right) which have become a characteristic feature of Chinese art (lower right)

arêtes. This must be one of the most difficult landscapes in the world to traverse – without a helicopter.

Yet another strange variety of tropical karst covers great areas of southern China. There, extensive sediment-floored plains are studded with isolated vertical-sided, limestone towers. They stand up to 650 feet high and most are riddled with caves. Chinese art has always been distinctive for the remarkable proportions of the mountains in any landscape paintings. Tall and narrow, they are not the romantic results of artistic license. Instead they are very true representations of one of the world's most extraordinary landscapes – a landscape almost unimaginable to one who has not seen it. In that way they much resemble those other remarkable features of karst – the caves.

3 Cave Exploration

'Because it is there' was the classic reply made by George Mallory when asked why he wanted to climb a mountain. Some cavers may give the same reply if asked why they want to go down a cave. Others might reply that they explore caves for the adventure, and some that they go for the physical challenge involved. Some go caving for the team spirit found in the sport, some go just to see what the underground world looks like, and a small minority might say they go down caves for the thrill and the danger. But, when posed the question 'Why go caving?', most cavers, and especially those who have been caving for a long time, would probably reply 'because I enjoy it'.

This last answer is the most difficult to explain. Cave exploration, caving, is a sport unlike any other. It involves some climbing and yet also perhaps some swimming, it is a personal challenge to anyone who does it and yet it also involves working in a team; and there is the added element that if due care is not taken it can be dangerous. Caving also takes people into a completely new environment, which can be very spectacular, but most of all, caving is different. It is so different that someone who has not actually been caving cannot really conceive what it is like. The whole idea, of people wandering about in little holes far below the ground, does at first appear a little bizarre; but it is really no more ludicrous than one man trying to kick a ball into a net while eleven other men try to stop him! A popular concept of caving is 'crawling about in muddy little tunnels'. Fortunately this is not true – for there are not many cavers who would enjoy just that. Walking, climbing, crawling, swimming and squeezing are how a caver moves about in his caves; and most of the distances are covered by walking. It is just that the climbs, crawls, swims and squeezes are more memorable and occupy a larger share of the time spent underground. Add to all this the darkness and the undeniable excitement of exploration, and one can start to understand the fascination which continually attracts cavers to their sport.

Squeezing through tight places is one aspect of caving which probably frightens a large proportion of people. Fortunately most cave squeezes are quite short, and widen out both sides into more roomy passages. A short squeeze is a challenge to the caver – a problem to be solved, particularly if there is a bend in it which involves fitting the body round in the one possible position. Long stretches of very narrow passage are not a feature of most caves – indeed there the challenge rapidly becomes mere hard work. The limit of exploration in many a cave is just a section of passage which is too narrow, so indeed to reach these limits very thin cavers have a distinct advantage! There are of course also psychological barriers to squeezing between walls of unyielding rock – so it is the thin confident caver who gains most ground in

In Iran's Ghar Parau (far right) the caver never has to squeeze but continually needs to contort himself down the narrow passage. A real squeeze has to be passed to get into Pippikin Hole (right) in the English Pennines. Not so tight but rather more interesting is the Slit Pot squeeze (below) which opens into the roof of an 80 foot shaft in the nearby Simpson's Pot

the squeezes. Two of the most successful thin cavers are the Brook brothers, Alan and Dave, whose home caving ground is the English Pennines. They have a whole list of explorations to their credit, where they pushed a little bit harder than anyone else in the squeezes, and so got through to discover new caves.

Marble Sink, a little cave on the slopes of Ingleborough Hill, was one of their more extraordinary explorations. Before they came across it, Marble Sink had been explored along a nasty little crawl and down a small shaft into a chamber with a slit in the floor dismissed as impossible – twisting downwards, it was less than seven inches wide. Then one day the Brooks went down the cave, and continued straight down the 'impossible' hole in the floor, to explore a series of passages and further squeezes which still constitute one of the most difficult caves in Britain. A number of other cavers have now been down the cave – since the Brooks removed the psychological barrier of the impossible slit – but it is still not a cave visited by large or even normal-sized cavers. Affectionately known as the 'Bastard Hole', that little slot in the chamber floor is in fact now about half an inch wider; one visitor got down through it, but decided he would never get up it, without gravity to help him; so he spent three hours hammering minute chips off the wall to ease his return journey to daylight. Less than a mile away is another of the brothers' discoveries, Hangman's Hole, where Dave Brook excelled himself in forcing a way through a squeeze. He pushed himself through an extremely tight vertical hole near the bottom of the cave, but, on his return, just could not fit himself back through it when he was having to push upwards, and therefore more strenuously. So his friends with him, who had not passed the hole, pulled him out with ropes. But even this was difficult, and only at the expense of the fracture of Dave's left collar bone!

It is perhaps surprising how few people do become stuck in squeezes in caves. Usually a pull or a push from a friend solves the problem, but it must be a traumatic experience to be stuck for a long time. The American caver Floyd Collins provided the almost unique tragedy of becoming helplessly stuck. Exploring Kentucky's Sand Cave in January 1925, he probed feet first down a little rift passage. He tried to kick a rock aside, but it rolled over onto his ankle and jammed; and he couldn't reach down to move it away. So his friends came to try to help him out. They gave him food and drink, and they tried everything – reaching past him, excavating a route around him, forcibly pulling him out, and eventually drilling a hole from the surface. But for seventeen heart-breaking days everything failed, and then he died. By that time a morbid crowd of more then twenty thousand had gathered at the entrance to hear the latest tragic news – and Floyd Collins had earned himself a place in history.

Looking down the rocky shaft of Eldon Hole in the English Peak District, the caver can just see the ladder shrinking into the distance

Large cave passages provide no such problems. Cavers often claim that walking down some mammoth tunnel provides no sport comparable to squeezing through tiny tubes, but few would not admit that they would prefer to discover passages of the larger variety. There is something quite special about walking five abreast through a huge cave, perhaps decorated with stalactites, and hundreds of feet below the ground. Unfortunately passages like this, with level sand or rock floors which permit such easy walking, are rare. Instead, boulders and breakdown piles are typical in the largest cave passages, and though travel five abreast may be possible, it is more usually by climbing and scrambling, not walking. Many of the huge French caves, the Berger and Pierre St. Martin included, have these boulder-strewn tunnels. A visitor to the Berger once made an apt analogy to 'ants in a sugar bowl', but he could well have improved this by mentioning the sugar lumps as well. Boulders fifty feet in diameter are not uncommon, and with the power of a caver's light lost in the massive tunnels, it has not been unknown for two groups of cavers to pass each other without realizing it – by going round opposite sides of a single boulder.

Probably what distinguishes caving most from mountaineering in the dark, is the vertical descents. Potholes and shafts, pitches as they are called by the caver, descend from the surface and from levels inside caves. Some caves start from the surface as horizontal openings, but inside have a series of ten, twenty or thirty pitches leading down to the depths, like a grand staircase. Pitches are climbed using ladders and ropes – and carrying this equipment into the cave is often the most arduous part of caving. So why not just climb the pitches by ordinary mountaineering techniques? On some pitches there are enough holds to make this possible, and then it is done; but in most cases the smooth, water-polished, vertical limestone walls offer no holds at all. Climbers could only surmount these by using tedious bolt-drilling techniques, and caves rarely offer an alternative route such as on a mountain face.

The basic item of modern caving equipment, next to the lamp, is therefore the wire ladder. With wires only a tenth of an inch thick, and rungs each six inches long at ten inch intervals, a convenient thirty foot length weighs only three or four pounds. And a whole series can be clipped together for a long pitch. The end-product is of course flexible, and swings around unnervingly on a large free-hanging pitch. Only a good technique avoids all the climber's weight being thrown on his arms, and a ladder pitch of perhaps two hundred feet can be impossible for an inexperienced caver. A nylon lifeline always provides the essential security, and indeed it can be an exhilarating experience to climb a long ladder. On the other hand, inexperienced or tired

Vertical caving in the English Pennines

In Hurnel Moss Pot (above) the ladder hangs down a finely fluted shaft, but in Lancaster Hole (far right above) it leans against a broken and jagged wall. In Little Hull Pot (right) the caver has to follow the waterfall on the ladder climb. The caver in Lost Johns Pothole (far right below) is not using a ladder as he prepares to climb the rope with mechanical clamps

cavers often find ladder-climbing a complete nightmare – especially on a difficult pitch, perhaps with a waterfall following the line of the ladder.

There was a time of course when wire ladders were unheard of. The earliest pothole descents, made over a century ago, mostly consisted of tying a man on the end of a rope and just lowering him into the depths, then to be pulled up again by a team of perhaps twelve or fifteen men. The great danger in this technique was that the man at the 'sharp end' of the operation had no control over his movements once he was out of shouting distance. Late in the nineteenth century a quarryman named Joe Plumley descended a hole in the English Mendip Hills by this method. Unfortunately the shaft he was lowered down was shaped like a bottle, and though he was lowered safely, on the ascent he caught himself beneath an overhang where the shaft narrowed. The team on top assumed he was slightly jammed in a narrowing of the shaft, so they just pulled harder – and broke Plumley's neck.

It was the great French caver Edouard Martel who, towards the end of the last century, revolutionized vertical caving by his masterful use of rope ladders. The world's first true professional caver, Martel toured France and then Europe and the United States, descending any number of open shafts for the first time ever. He even beat the fairly active British cavers of the time to the first descent of England's greatest shaft – the 360 feet deep Gaping Gill. Back in France he also explored the 550 feet deep shaft of Jean Nouveau on the Vaucluse Plateau. It is often claimed that Martel relied heavily on a strong pulling team on his lifeline, and only really used the ladder to control his climbing. Nevertheless his feat at Jean Nouveau was quite remarkable at the time, and to this day a depth somewhere around 600 feet is still reckoned to be the maximum for a ladder climb.

Martel's use of ladders was really exceptional. Few other cavers ever climbed such long ladder pitches, and even up till the 1960s the discovery of a shaft more than about 500 feet deep was normally the sign to obtain a powered winch. Indeed many great shafts, particularly in France, were explored by the use of winches. With a surface pitch such as the 1050 feet deep Lepineux shaft in the Pyrénées, the technical problems were not too great – though accidents did happen, and the famous French caver Marcel Loubens lost his life on this very exploration. Most difficult were the underground shafts – such as in the Henne Morte pothole, also in the French Pyrénées, where a winch was assembled at a depth of over 600 feet to descend a further waterfall pitch, itself more than 300 feet deep. Some sort of telephone was always used to enable the winch controller to liaise with the 'victim' in the hot seat, to sort out problems as they occurred – but even this had problems.

One of the last great winch explorations was that of Provetina – a hole in the Pindus Mountains of northern Greece. Found by English cavers in 1965 it was merely known as 'very deep'. Parties went out to the shaft in 1966 and 1967 but could not get beyond a steep avalanching snow ledge 570 feet down. Then in 1968 two separate expeditions went to Provetina; both were rather small and inadequately equipped. The first group to arrive managed more by luck than anything else to reach the bottom, at a level 1285 feet vertically below the entrance – the world's greatest known shaft at the time. The second group then arrived, and heard they'd been beaten to it, but three of them still decided to 'have a look'. They took their winch up the mountain, assembled it on the edge of the shaft, and Pete Livesey, one of Britain's finest cavers and climbers, strapped himself into the hot seat.

Slowly he descended. He had a radio to talk to the winch operator, and as he went down he gave a running commentary about his exciting journey. He passed the snow ledge, but beyond there, unknown to him, the radio signals did not get back to the winch, and he just continued talking into the microphone. Up on top they didn't bother about the lack of radio, as they knew there were no more ledges or difficulties, and they just stopped winding when they knew he would be at the bottom. They then gave him five minutes to look around and started to winch him back up. Meanwhile Pete had assumed they had stopped winching him down because he had told them to, and he'd left the winch seat for a look around the roomy chamber in which he had landed. He was on the far side, taking photographs, when the seat started to go back up again. He ran across, caught it with one hand, and slowly rose up into space! 'To go or not to go' was the question, and realizing he'd never hold on all the way up, he rapidly let go and fell back to the floor.

When the two on top saw the winch seat rise out of the gloom empty, they guessed what had happened. But every time they sent the seat back down it caught on the snow ledge. Eventually one of them, Sion O'Niel, went down to the snow ledge himself. There he took up a precarious stance tied to some loose flakes of rock, and threw the winch seat over the edge on its journey downwards. A seemingly endless wait later Pete came up over the edge, and it wasn't then long before all three were back together again on the surface.

Nearly all the basic caving techniques and apparatus have originated in Europe, but recently the Americans have made a great contribution with their use of single ropes without ladders. This does not involve hand-over-hand rope climbing – an exercise normally fatal on any pitch longer then ten or fifteen feet. Instead it involves abseiling down ropes – sliding down on mechanical braking devices – and the ascent depends on Jumars or

Maypoling a waterfall shaft in Langstroth Pot, England

some other similar form of ratchet clamps which grip the rope. The American cavers have found long dry pitches to be much easier when climbed on ropes, and the practice has now spread around the caving world, though shorter or wet pitches are still best tackled with ladders. The Texan cavers really developed their rope techniques when they started the exploration of the big shafts of the Mexican caves. Their great test was Golondrinas – a vast pit descended by a single unbroken free-hanging pitch of 1092 feet. The ascent of this spectacular shaft takes most cavers over an hour, so there is good reason to perfect their techniques; the 'ropewalker' method seems the best, using a clamp on each foot to walk up the rope and another clamp on a seat harness for sitting back for a rest. One caver even invented 'MAD' – a motorized ascending device. A two stroke motor took the hard work out of rope climbing, unless – and it actually happened to one caver – one ran out of petrol.

Exploring a cave upwards is doing it the difficult way, but sometimes it is the only way. Shaft climbing then involves time-consuming acrobatics drilling bolt holes and using 'maypoles' – scaffolding poles up to fifty feet long used to push wire ladders up pitches. Considerable climbing skill is also a help on the more broken shafts. Of the world's various upward cave explorations, that of the Trou de Glaz in the French Alps must surely rank as the finest. The Glaz entrance lies in the west flank of the Dent de Crolles massif, and from 1941 to 1947 Pierre Chevalier and his team visited the cave 31 times trying to find a way between there and the shafts on the plateau a thousand feet above. All the shafts were blocked, so the team had no alternative but to work upwards from the Glaz; they climbed and maypoled more than 600 feet of pitches and explored nearly a mile of ascending passage until they could see daylight through a tiny crack in the roof of their cave. Blasting the fissure open they had a route into the base of one of the plateau shafts and they had successfully explored one of the world's finest caves. Though it cost Chevalier's team seven years' hard work, cavers today descend the cave in less than ten hours.

When a modern caver goes underground, his visit rarely lasts more than about twelve hours. So a miner's electric cap-lamp can provide him with a good reliable light: it is waterproof, and is only really inadequate in the very largest of boulder-strewn passages. With the lamp on his helmet, turning with his head, the caver barely notices the darkness. If he has to use an acetylene lamp, with its naked flame light, he can still see well, but may notice some inconvenience if he walks under a waterfall and it goes out! But he will barely spare a thought for the pioneers of 70 years ago, who explored so many great caves using just candles to see their way. Since a waterfall pitch in a cave is best

The mud wallows in Pippikin Hole (above right) in the English Pennines provide cavers with a rather odd type of enjoyment. Nearby, the Gaping Gill cave involves a different kind of challenge; this vertical squeeze (below right) is at the top of a ladder pitch – so there is 70 feet of space beneath the caver's feet. A narrow ledge provides an exposed route along the chamber wall in Belgium's Père Nöel cave (left)

To each cave its own obstacles

Three caves in the English Pennines each provide their special problems: how to pass the squeeze in Pippikin Hole without winning a muddy face (far left); how to traverse round the pool in Lancaster Hole without getting wet above the legs (left); how to climb down the cascade in Goyden without having a shower (below). In the larger passage of the Berger Cave in France, the traverse round the pool (below left) avoids a full-scale swim

likened to a thunderstorm and gale on a dark night, the problems with candles seem almost infinite. Indeed, they were hard men who were prepared to be lowered by their friends down a waterfall clutching a candle in a jam jar. Stories abound of the difficulties the early cavers had when their lights were extinguished.

Without a light, a caver would be helpless in all but the simplest of caves – he would be hopelessly lost. There are just a few cases known where people have wandered into caves, lost their lights, got lost themselves, and eventually died. But properly equipped cavers almost invariably find their way out again safely; at the very worst their friends have to go down to look for them and help them out. Some caves make the famous Hampton Court Maze look simple, with their endless numbers of side passages. Explorations in this type of cave can be quite interesting. A journey to the far reaches of Mossdale Caverns, in the English Pennines, involves about half a mile of crawling-size passages. They were first visited by an exceptionally tough caver, Bob Leakey, during a magnificent solo exploration in 1941, and the next people to go there were Pete Livesey and Mike Boon in 1964. They compared notes with Bob Leakey, but could not agree on the description of the last half of the passage. The problem was only solved in 1966 by two groups of student cavers from Leeds and London. The London group followed Leakey's description and found their way to the end of the long crawls. But on the way back, they got lost, and took a wrong turning. Nearly a thousand feet of crawling later, a tight squeeze made them realize their error – but then they recognized where they were, back on the right route. They had returned by way of a route first explored by Livesey and Boon – there were two ways through this incredible system of crawls, aptly named the Marathon Passages.

There are two ways of laddering the Wet Pitch in the Pennine cave of Lost Johns (left) – short and wet down the waterfall, or long and dry from a ledge high in the roof. The long way has the advantage of being possible in flood conditions (above) when anyone would be swept off a ladder hung down the waterfall

Bob Leakey first found his way into the Marathon Passages by following a tiny stream. The stream caves are always the main attraction for explorers. Streams form the caves, they are the arteries of the cave systems; active, noisy and alive they contrast sharply with the old, silent and dead, dry caves. Cavers always follow them if they can – and of course downstream is usually the quickest way to the depths. To the exploring caver, a fabulous stalactite grotto may be a very welcome bonus, but the ultimate aim is always the discovery of a long streamway. Best of all is a complete through-exploration – following a stream down a sinkhole cave, and emerging hundreds of feet lower down, with the stream, from its resurgence cave. But rarely is this ideal attainable, for cave streamway exploration is not without its problems.

Following a river or stream in open air is simple, by walking along the banks. But a cave stream commonly occupies the full

D

width of the passage, so following it means walking in it, wading through pools, swimming through lakes and climbing down waterfalls. The latter are the greatest obstacles, for fast moving water presents a force far stronger than man, and a caver could easily be swept off a ladder by an oversize waterfall. The great Berger cave in France consists essentially of a single stream-passage, nearly three miles long and descending over three thousand feet. Fortunately though, the stream is quite small, and for most of the route the caver keeps dry, walking over boulder piles and in higher levels of the passage. But for two stretches there is no way except in the water – swimming, wading, clutching for handholds along the walls, and having the greatest problems on the waterfalls. From the lips of the larger falls the caver has to climb out along the walls till he can hang a ladder clear of the main cascade – and climb down merely under the soaking spray.

The Berger stream only provides real problems on the waterfalls, but a much larger cave river can be far more troublesome.

Paddling down the streamways is one of the delights of caving, in the Kingsdale Master cave (far left) and in Lancaster Hole (below left) both in the English Pennines. But in Ogof Hesp Alyn cave (left) in north Wales the caver gets more involved with the water

The down-stream end of Ethiopia's Sof Omar Cave contains a maze of passages where the limestone has been eaten out along the joints

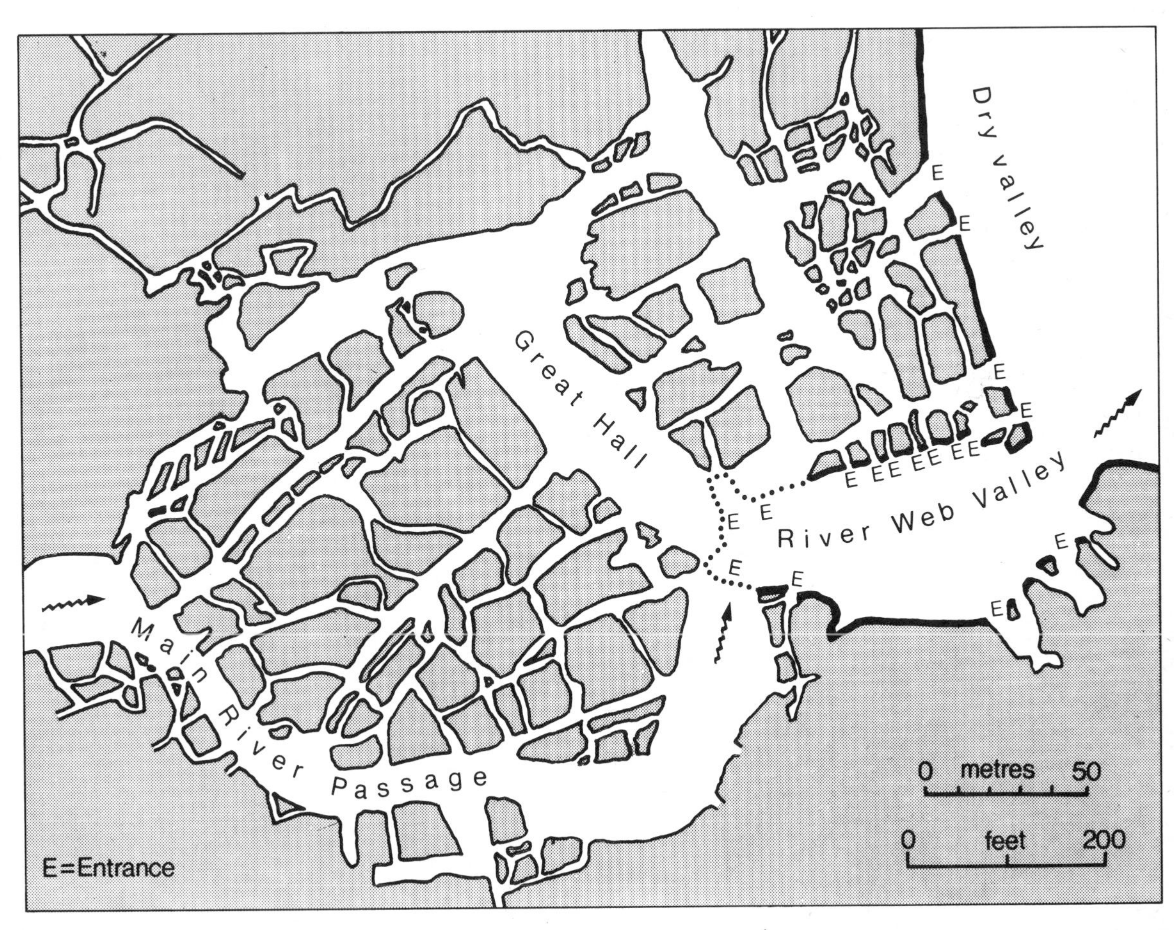

Wet caves have their own special appeal – the Han Cave in Belgium (below), and Pippikin Hole in the English Pennines (centre and right)

In western Guatemala a huge river flows through the cave of Agua Escondida. This has been explored in an upstream direction for over two miles, and though the passage averages fifty to a hundred feet high and wide for much of its length the river occupies the entire floor-width. Wading and swimming against the current in the deep parts is difficult enough, but where the river flows down rapids progress is only possible by such techniques as boulder-hopping or climbing and traversing along the walls. One of the most difficult sections of the cave is where the passage narrows to twenty feet in width, with walls plunging straight down into the fast flowing water. The explorers only forced a route through by sitting in their rubber dinghy and pulling themselves along using handholds on the wall. Further upstream a three foot high waterfall provided a major obstacle with

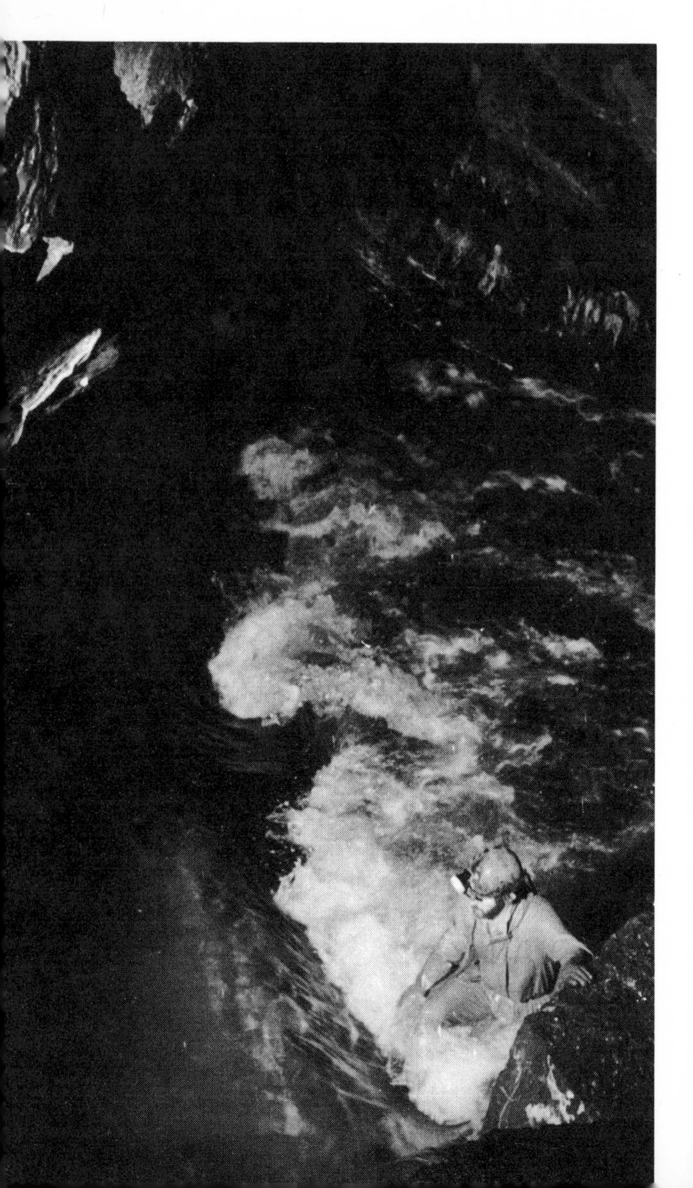

its great masses of white water just above and below. Beyond this a similarly turbulent six feet cascade stopped the original explorers. This obstacle was only passed in 1973 by that fine ex-patriate English caver, Mike Boon – and he had to resort to throwing a lassoo over a projection in midstream and hauling himself up on this. He then explored the cave onward to a sump; the return journey, downstream, was a pleasure – jump in and be washed along by the current.

One of the most enjoyable sides of caving is sitting in a rubber dinghy and gently paddling across a calm lake in a beautifully decorated gallery. Underground boating means that the cave is a true 'water cave', and those are the best. Inflatable dinghies are indeed standard equipment for many caves, though unfortunately none in Britain. A caver can always swim across a small

The top of the cascade brings an escape from the water in Wales's Ogof Ffynnon Ddu

(Overleaf) rubber dinghies give the cavers an easy ride along flooded tunnels of the Krizna Jama cave in Jugoslavia

lake, but there is a limit to how far he can go if he has to carry ladders and ropes to equip further pitches. Unless there are plenty of holds on the walls which can be used for support, a dinghy may be essential for this reason alone. One of Europe's finest boating caves is the Krizna Jama in Jugoslavia. The cave is entered near the resurgence, and is explored upstream, and there are no pitches requiring equipment to encumber the visitor. A succession of twenty-two lakes makes the journey a delight. Rubber or wooden boats are used, and nearly all the time is spent in them – only short portages are involved, where the boats have to be carried through dry passages or over boulder piles. For its whole length the cave averages thirty feet high and wide, and there is never enough flow on the lakes to make progress difficult. Some of the tunnels are lined by stark black rock, and others have fabulous collections of stalagmites and stalactites. No one can fail to thoroughly enjoy the Krizna Jama.

Unfortunately not all cave boating is quite as pleasant as that. The Padirac Cave in the Dordogne region of France has a terrible reputation. The entrance series is a very fine show cave, where a wide 300 foot deep shaft drops into a tall canyon passage. Tourists finish their visit with a boat ride on a large clean lake. Beyond is the world of the cavers only, and though more than three miles of passage have been explored there have been few visits to the far reaches. The cave has little gradient and only quite a small stream, but this passes through a whole series of lakes and beneath intervening boulder piles. A visit to the end takes so long that an underground camp is needed, which means that the caver is encumbered with extra equipment; and in addition the cave is renowned for its mud. A trip into the Padirac cave is therefore an endless succession of struggles through deep liquid mud, short boat journeys in overloaded rubber dinghies, and tedious and difficult climbs over boulder piles, laden down with the boats and equipment. It is wet, cold, dirty and miserable – essentially everything that is bad and untypical about a cave – but of course it is a perpetual challenge.

Despite the strong attraction that always draws cavers to them, underground streams and rivers do present their own problems. Particularly in a mountain region, the levels of streams can change very rapidly in response to weather variations. It is the unpredictability of weather which makes water levels and floods the least predictable feature of caves. There are many times when experienced cavers would not dream of going down certain caves, because there is too much rain falling or forecasted. But in the climate which affects most of Europe, and many other caving areas, rain is not always foreseeable. Most seasoned cavers have, at some time, experienced a rapid increase in water flow when underground. In some cases they have re-

treated to a dry, safe chamber to wait for the water level to drop again, but in most cases a flood just makes the journey more exciting. Indeed, most caving regions have caves which are known to be safe during even the highest floods. Some are abandoned and permanently dry, but others can take an immensely increased flow of water and still offer a safe route for cavers. Local cavers often visit some of these known safe caves in wet weather, just to experience the raw power of a foaming brown flood roaring through its underground channels.

Besides all the excitement and spectacle created by a flood in a cave, there are also problems, particularly if the cave cannot really cope with the flood. France's Berger cave is one that rapidly becomes impassable during flooding, and as a full descent of the cave takes a few days, it is all too easy to be out of touch with surface conditions and get caught out by a flood. In 1964 a strong British expedition had established their second underground camp in the Hall of the Eight, 2600 feet below the surface. This campsite is dry and safe in all conditions, but the day after setting it up a small group set off downwards again, placing their ladders on the lower series of pitches. While they were down, it rained hard, and the cave flooded. On the hundred feet deep Grand Cascade, just below the camp chamber, the newly grown waterfall washed over the ladder and made the pitch unclimbable. The group below retreated to a dry chamber and prepared to wait. They had lights and some food, but only the wet clothes they were wearing. They could not risk getting too cold, so they dare not sleep for too long at a time. Every hour they got up, jumped around and ran on the spot, till they were warm again; then while most sank into a shallow shivering sleep one stayed up to wake them again for their hourly exercises. They did this for two whole days and nights, till the flood abated, and they could rejoin their friends in the relative comfort and sleeping bags of the camp just a hundred yards further up the cave. The whole episode was written down to experience, and they all continued caving for many years later.

Unfortunately two days is not the longest that cavers have been trapped underground by floods. In 1952 Alfred Bögli and a party of three young cavers were exploring the Hölloch – a huge cave in the Swiss Alps containing more than fifty miles of known passage. Nearly all of this is dry and safe in any weather, but the cave has only one point of entry – at its lower end. Between the entrance and the main cave the passage dips down in a huge U-tube and this can fill with water to become completely impassable. When Bögli's party found they were cut off they retreated to a large dry sand-floored chamber and made themselves comfortable. Some of them went to various points in the cave to gather up food and lights from emergency dumps which had been

Caves in flood

In Marble Steps Pot in the English Pennines (left) flooding sends a cascade down a normally dry shaft. And in the nearby Lancaster Hole (above and below) the cave stream can turn into a foaming torrent

The Aiguilles cave system descends in a series of steps through the limestone mountain at Devoluy in France

established by the cavers, ready for just such an event. They were cold in their cave, but they were dry, and the main difficulties involved in their stay were psychological. But they talked and occupied themselves, and made periodic visits to inspect the water level. Eventually, after ten days, the water drained away, and the cavers emerged safe and none the worse for their experience.

Water in caves can of course also present a much more absolute danger. There is no denying that caving is one of the sports which involves some risks; but care and forethought can eliminate most of the dangers. Similarly care can prevent nearly all the perils of flooding, but just occasionally the combination of a freak rainstorm and a flood-prone cave can lead to tragedy. Indeed the world's two worst caving accidents have both been due to flooding. In June 1967 a team of six experienced cavers visited the far reaches of Mossdale Caverns in the English Pennines. This involved crawling down the half-mile of Marathon Passage which is rarely more than a couple of feet high. While they were down, an exceptionally heavy summer thunderstorm hit the area. The river entering Mossdale rose to a roaring torrent, and its passage inside the cave rapidly became completely full. Over-

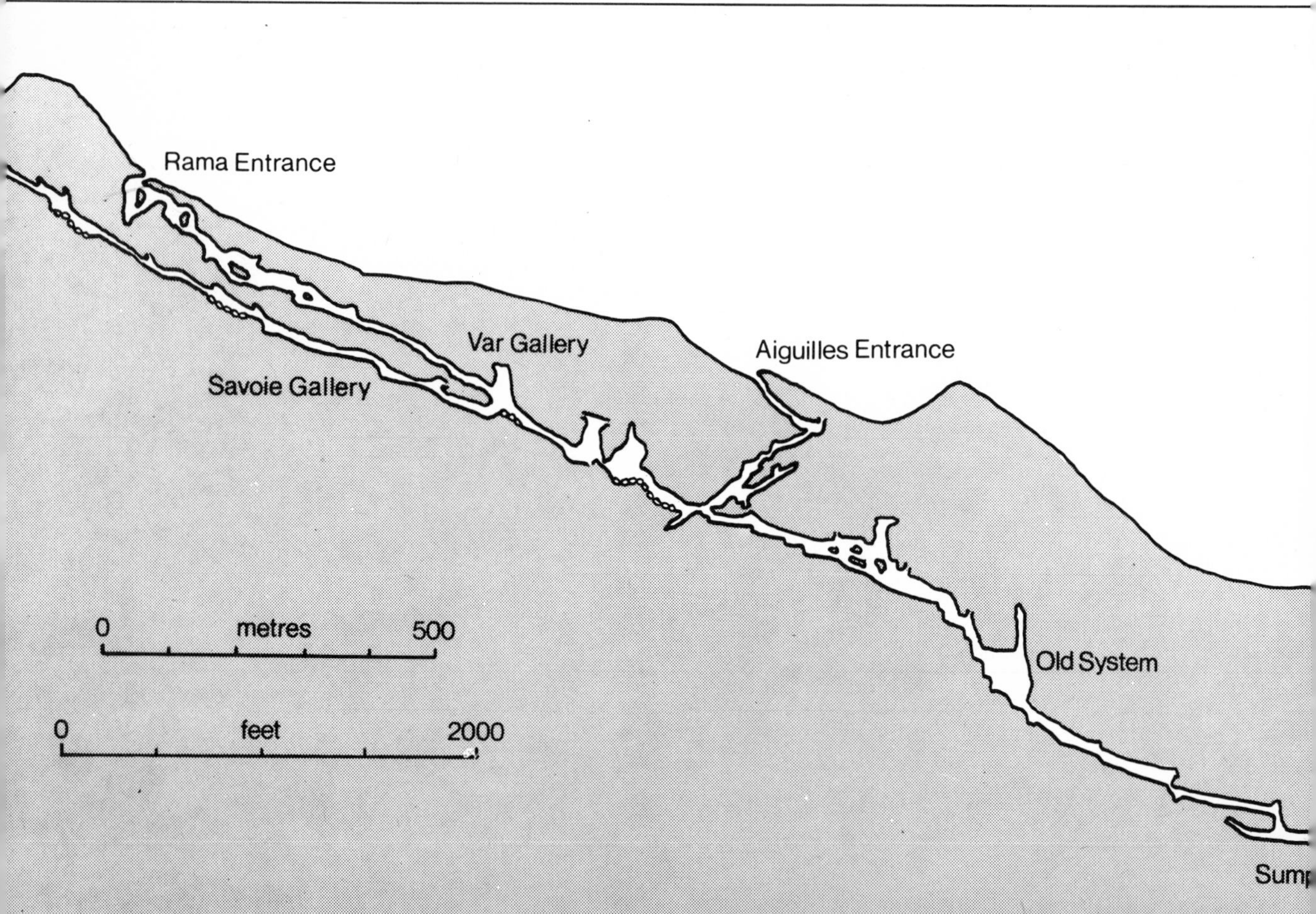

flow water in vast quantities therefore ran off down another passage – the one to the far reaches, which normally carries only a tiny, separate stream. The flood wave met the cavers part way along the Marathon crawls, and they stood no chance; all six were drowned.

The great tragedy of the Mossdale accident was that, experienced though they were, the cavers could do nothing to help themselves. In contrast, the other great flooding accident was largely due to panic setting in with inexperienced cavers. In November 1950, a party of seven cavers were visiting the resurgence cave known as the Trou de la Creuze in the French Jura. A freak flood made the cave river rise dramatically, and the party started to beat a hasty retreat downstream towards the entrance. But they soon met a flooded section of passage where the water met the roof. One after the other the cavers dived into the flooded section. The very experienced leader of the party, Dr. Mairey, arrived at the point last – he had been exercising a little more caution as he moved down the river passage. He called to his colleagues not to dive, but they ignored him. He did not know how long the flooded section was, and he considered it safer to wait at a spot where he thought the water would never rise to the roof. He stood on a ledge as the water rose steadily, and then his light faded out leaving him in inky darkness. For twenty-seven hours he stood in the shoulder deep water, before the flood abated, and caving friends made their way up to him. Dr. Mairey saved himself by thinking sensibly in a difficult situation. All six of his team-mates who dived into the flooded passage were drowned before they could reach the next airspace.

Similar clear thinking saved the lives of five French and Spanish cavers exploring the great cave of Pierre St. Martin in the Pyrénées during 1966. They were the second party ever to explore the upstream section of the great river gallery of this magnificent cave. A good campsite in the last of the dry chambers meant that they could set off into the river early in the morning. But it was in full flood. For two days they sat in their camp waiting for the water to go down, until eventually they could set off upstream. The first quarter of a mile was easy, but then they had to resort to their rubber dinghies. Deep pools made progress slow, especially for the one member of the team who did not have a frogman's suit; Isaac Santesteban only had chest-high waders to keep him dry, and of course they were useless for swimming. Two of the team went ahead with one boat, the other three, including Santesteban, progressed more slowly with the second boat. Then their boat capsized and Santesteban was soaked to the skin. He got on to a ledge, stripped his clothes off, wrung them a bit drier and dressed again. Meanwhile the water level rose – a second flood was coming down the cave.

They had to go on, to meet up with their friends out in front. For hours they struggled against the torrent, until further progress was impossibly dangerous. They climbed onto a ledge and sat down; Santesteban was now really cold and his two colleagues took it in turns to rub him to restore his circulation. Hours later they heard voices and were staggered to see their two friends approaching them – bobbing around in their dinghy on the foaming torrent with their lights absolutely dead. They hauled them, dazed and frozen, on to their ledge. The water continued to rise till even the ledge was submerged, and Santesteban was growing rapidly worse. With no alternative available, they lowered a dinghy onto the torrent, installed two of them in it, and allowed them to be carried downstream on the end of a rope. Twenty yards on they climbed out on to another ledge. The dinghy was hauled back, and two more cavers repeated the process. And so it went on from ledge to ledge down this foaming underground river. Progress was painfully slow, but any other method would have been foolishly dangerous; so they persevered. And eventually they reached the haven of warmth and rest offered by their sleeping bags in their dry campsite. Their struggle against the floodwaters had lasted 55 hours, but they were safe, and had some incredible memories to look back on.

A cave passage permanently flooded to the roof is almost always an impassable barrier to a caver's progress. However there are, in many countries, small groups of cave divers; and the flooded caves are their world. Cave diving is not only a unique, specialist sport, but it is also dangerous – combining all the hazards of both caving and diving. Standard equipment includes a frogman's neoprene suit and a 'scuba' diving set, together with helmet and lamp, boots not flippers, and a line reel; nil visibility is common in muddy water and a cave diver must always lay a line to provide a route back. The unique thing about cave diving is that a diver in trouble cannot just rise to the safety of the water surface – there is no easy way out of a flooded cave; so he is totally reliant on his equipment, and regards a 100% safety margin as an absolute necessity. Compensation for these difficulties and dangers are both the privilege of exploring the silent world of the flooded caves, and also the possibility of discovering new 'dry' caves beyond the flooded sections.

Cavers know the flooded sections of caves as sumps, and every diver aims to pass the sump into which he plunges. Failures are all too common – even the slightest of squeezes can stop a heavily equipped diver. Furthermore it takes a lot of stamina and courage to dive into a very long sump. In both Italy and Florida single dives of over 3000 feet have been recorded, but these were exceptional. And even the end of the sump is not always the end of the difficulties.

Coming up into an airspace in the flooded section of England's Wookey Hole, the cave diver clings on to his reel of line which will lead him back through the inky waters

Quite close to each other in the French Vercours are two sumps of very different character. The Cholet cave is a resurgence cave in a magnificent setting at the foot of the Combe Laval – a thousand foot high amphitheatre of vertical limestone cliffs. Just inside the Cholet, the cave sumps, and some years ago it was dived by a caver from Lyons. After a dive of 500 feet he emerged in a pool – with a waterfall beating down on his head. The walls rose vertically on all sides and there was no escape from the beating water – except a retreat. Luck was better at the nearby Coufin spring, where a large stream flowed out of a pool at the foot of another great limestone cliff. There the diver had to swim less than a hundred feet before emerging in a fabulous cave lake. Thousands of pure white stalactites hung from the roof, and two beautifully decorated stream passages entered the far side of the chamber. It was a magnificent discovery, and the subsequent removal of the sump, by lowering the level of the spring pool, made it accessible to the non-diving majority of cavers.

A cavér climbs the swinging wire ladder just out of reach of the 150 foot high waterfall in Juniper Gulf in the English Pennines

Swildon's Hole is one of the finest caves in the English Mendips. In 1921 it was explored as far as a sump only half a mile from the entrance, but it was not until fifteen years later that the sump was passed – using an extremely crude home-made breathing apparatus. That first sump is less than six feet long, but only a thousand feet further down the streamway Sump 2 proved a longer and more intractable obstacle – impossible with the equipment of 1936. But techniques improve and attitudes change, and by 1950 further generations of cave divers had explored down the exciting stream cave to the sixth sump. The seventh sump was reached in 1961 and the following year this was only passed after a terrifying struggle by the first-class diver, Mike Boon. The sump is only a few yards long but contains two difficult squeezes. Mike had to take his bottle off his belt and hold it in his hand to pass the squeezes, and even then he nearly got stuck underwater. It was three years before anyone again approached Sump 7, and then they did so in order to place an explosive charge to remove the worst underwater squeeze. Now exploration in Swildon's has reached the twelfth sump.

Once a sump has been passed by a fully equipped diver, the danger of the unknown is removed. If the underwater passage is straight forward and not longer than about thirty feet, it may then be free-dived – passed by a caver without breathing apparatus. Sumping is yet another exciting aspect of caving, and a trip down to Sump 9 in Swildon's Hole is a classic. Unfortunately Sump 9 is over a hundred feet long, so beyond there the cave is still exclusive to the cave-divers.

No one would intentionally free-dive a 100 foot sump – though the record stands at a full 120 feet! This was achieved by a sixteen-year-old army cadet John Stevens in 1965. One of a party of cadets lost in Carlswark Cave in England, he dived into a flooded passage in his search for the way out of a cave. He did not even have a waterproof lamp and he swam the whole 120 feet in darkness to emerge in a tiny airspace – where he then stood on a boulder shivering with cold, and too terrified to go onwards or back. Meanwhile the others of his party found the right way out of the cave, and went for help. The cave rescue teams – made up of all the local experienced cavers – turned up; and they felt sure they only had to look for a body. A cave diver went into the sump, and explored it to an underwater heap of boulders at the end, finding no sign of Stevens. He had not even seen the side turning which, by an extraordinary chance the cadet had followed to the one and only airspace! So the rescue teams decided to lower the water level; after a whole night's pumping there was an airspace right through the passage – and they could hear young John Stevens whistling at the other end. It wasn't long before he was out, and surely one of the luckiest people in the world.

The Well Pitch in Ireby Cavern (far left) in the English Pennines is an easy climb with the ladder well clear of the waterfall. More difficult is the climb straight from the rubber dinghy onto the ladder in the Gournier Cave (above) in France. The Gournier lake (left) can only be crossed by boat or swimming

John Stevens' mammoth free dive was, perhaps fortunately, unique, but almost as remarkable have been the great submarine cave dives in the Andros Island region of the Bahamas. All around Andros the clear warm coral sea is no more than about twenty feet deep, except where the sea-bed is punctuated by the Blue Holes. These are immense flooded potholes, mostly about a hundred yards across, with vertical walls plunging to depths of up to 300 feet. Proof of the fact that they were formed on dry land as normal cave systems is furnished by the barnacle covered stalactites hanging down the walls; and, furthermore, cave passages open from their walls. The famous French marine scientist, Jacques Cousteau, has explored some of these submerged caves. Armed with everything from film cameras and lights to midget submarines, he and his team of divers swam through these caves which have been flooded for perhaps 10,000 years. Their finest exploration took them to a depth of 150 feet, and then along 800 feet of tall narrow cave into a beautiful chamber decorated with slender stalagmites ten feet and more high. Spiralling around in this eerie green chamber, the divers experienced one of the most beautiful methods of cave exploration – floating through water is so much easier than scrambling over rock piles.

Really long caves may take more than a single day to visit, and can therefore require underground camps. Establishing camp in a cave is not a popular task with cavers, because of the amount of equipment that must be carried in. Food and sleeping bags are the two essentials, together with an airbed (unless a particularly sandy-floored chamber is available) and dry clothes if the journey down the cave is a wet one. Tents are of course superfluous – at least it doesn't rain underground. There is something peculiar about sleeping underground. Many cavers like it, because it eliminates the walk to the entrance to start the day's caving. But it is waking up into absolute blackness which is most odd; first man awake has to grope in the dark for his lamp – ideally prepared the night before and placed just by his head – before he can see or do anything. The next task is usually to prepare a hot drink. The Hall of the Thirteen, a fabulously decorated chamber in the Berger cave in France, is one of the most frequently used underground campsites. There, the first man up has to slip his feet into his boots and, lamp in hand, wend his way down the chamber between the magnificent stalagmites, to where he can scoop a panful of water from one of the beautiful lakes – it's a unique and unbeatable way of waking up to a new day's caving.

Not all campsites are as idyllic as the Hall of the Thirteen. Mullamullang Cave lies beneath the Nullarbor desert of Australia. It contains many miles of huge passages – but there is no

The 100 feet of ladder in England's Rift Pot takes the caver straight from the windswept Pennine fells into the boulder-strewn chamber where the only sound is that of dripping water

Ploughing through thick saturated mud is the only way along Pippikin Hole in the English Pennines (right), while not far away there is no mud but plenty of water in Out Sleets Beck Pot (below). Naciemento del Rio Cave (overleaf) in the Spanish Cantabrians contains water deep enough to involve long spells of swimming

water in it. The 1963 expedition to the cave established an underground camp over two miles from the entrance, and, when the main difficulties in the cave were due to heat exhaustion, it was frustrating indeed to have to carry all the drinking water in from daylight. In 1954, the large expedition to the Flint Ridge Cave in the U.S.A. did have water at their underground camp, but all their equipment had to be laboriously dragged through a crawl passage less than a foot high in parts. But transporting equipment is not the only problem in cave camping. Winter 1973 saw a large group camped in Yorkshire Pot in the Canadian Rockies. Mike Boon was in the party, and at midnight on the second night down he woke up when all the others were still sleeping. Needing to answer a call of nature, he took a light and walked two hundred feet down the sand-floored passage to the chosen spot. Then on the way back he inadvertently extinguished his light; and it took him three hours to find his way back to his sleeping bag!

Treading new ground, exploring for the first time, and discovering new caves provide most cavers with their happiest and most memorable times. It is a fabulous experience to explore a new cave, but it is rarely an experience easily won. Walking over the surface of limestone areas, looking for where streams sink underground, is one way of finding caves, but in many parts of the world someone else will have been there first. Finding new caves requires considerable diligence. In Britain, the first part of the century was the golden age of cave exploration. Modern cavers can only envy the first people who walked up to Rowten Pot in the Pennines and followed its stream into the underground. Rowten is a wide open pothole a hundred feet long and thirty wide, with a clear stream cascading down a series of clean washed shafts to a depth of over 350 feet. Exploration must have been pure delight, unhindered by any constrictions or difficult sections. Seventy years after Rowten Pot was first explored, a complete contrast was provided by the discovery of Strans Gill Pot, still in the Pennines. Harry Long, one of Britain's most successful cave explorers, was examining the stream-bed of Strans Gill in 1967 when he noticed a little whirlpool where some of the water was sinking into an opening in the ground less than an inch across. He poked a crowbar into the hole, enlarged it a little, and saw that more gravel and water disappeared. A few minutes work and he could see a narrow rift in the limestone, but it took a lot more digging and rock excavation before the opening was large enough to enter. Eventually it was, and this forty-feet deep shaft, still only seven and a half inches wide in parts, is now the way in to the shafts, galleries and beautifully decorated chambers of Strans Gill Pot.

Even worse to open up and explore are the caves with their

entrances blocked by sediment and rockfall. These really try the patience and enthusiasm of the cavers. Rhino Rift in the English Mendips was an obvious entrance, situated in a dry valley, and clearly a major stream must once have sunk into it. But in the present climate and situation it is dry and its entrance has been filled with silt and gravel. The local cavers thought the situation promised well for a major cave system, so they proceeded to dig the sediment out of the entrance. Unfortunately there was a lot of sediment there; as some cavers lost interest others joined, and only after 25 years of sporadic weekend digging, did they finally break through into open cave and a series of shafts leading to a depth of over three hundred feet.

Rhino Rift was rather a disappointment to its discoverers. Though deep, it had little horizontal passage in it, and many miles of cave must lie beyond the impassable boulder choke which is the present limit of exploration. In this respect caving is full of disappointments – because it is almost impossible to foretell what lies ahead. In 1971 a party of English cavers went out to Iran to look for caves in the limestones of the Zagros Mountains; a small group, with relatively limited amounts of equipment, their main aim was reconnaissance. But high on the mountains they found the entrance to Ghar Parau – a fine cave containing a tiny stream which plunged steeply downwards. They explored it till they ran out of ropes after descending a succession of more than twenty shafts. Nearly a mile in and 2400 feet below the entrance the exploring team had to turn back at the top of a twenty foot shaft – looking down into a passage which curved into the unknown. Since the water emerged from a spring over 5000 feet below the entrance, there was still a great potential for discovering new cave. They vowed to return. 1972 saw a second stronger expedition to Ghar Parau – sixteen cavers, masses of equipment, and everything for an underground camp so that they could push on further into the mountain. The cave had every chance of breaking the world depth record. Down they went to the 1971 limit, and then down the pitch and into the unknown. Just a hundred yards further on, the walls closed in and the roof dropped down into a pool. The water disappeared into this narrow muddy sump, and there was no possible way through for the cavers. It was a classic disappointment – a whole expedition for just a hundred yards of cave! But, after all, caves are unpredictable.

Fortunately not all cave explorations turn out like that in Ghar Parau. One of the finest success stories is set in the limestones of Kentucky, U.S.A. The immense galleries of Mammoth Cave had been fairly thoroughly explored by 1950, and over forty miles of passage were known. The whole system lies under a low ridge, and to the north this is bounded by the Houchins Valley, with Flint Ridge on the other side. Even then a number of caves were known in the limestones of Flint Ridge, but it was only in 1954 that a bout of successful explorations revealed how extensive these were. Enthusiasm rose among the cavers when they realized the potential, and by the end of the next year most of the caves had been linked up to form one great network – exploration had shown that the short caves previously known were really just fragments of one. Then called the Flint Ridge System, the known extent of the cave continued to grow as exploration slowly progressed along the smaller, or more awkward, or more remote passages; and still the cavers were breaking out into vast new caverns – they were not just poking about in little crawlways. By the time the explored length of Flint Ridge cave exceeded that of Mammoth Cave, there was talk of a link between the two. The problem was to pass beneath the Houchins Valley – to find a cave in the relatively thin layer of mainly waterlogged limestone

In Giant's Hole in the English Peak District a series of cascades in the lower streamway provides plenty of amusement for visiting cavers

still present beneath the valley floor. The cavers kept on searching and probing in the passages nearest to the Houchins Valley, but it was only in September 1972 that they met with success. An active stream passage was followed downstream from Flint Ridge Cave. It was in a remote part of the system and even to reach its start involved hours of walking and scrambling through already known cave passages. Once inside, the water was deep and cold, but the explorers pushed on till they were wading through shoulder deep water. The roof became lower, until there was less than six inches of airspace. Heads tipped to one side, and faces half in the water, they kept going – until suddenly the roof rose. A high gallery stretched in front of them and they recognized it as a part of Mammoth Cave – the link was made. The known cave passages in the Flint-Mammoth Cave System now exceed 140 miles in length – a fine record of the enthusiasm generated by caving.

4 The Use of Caves

Caves constitute a small but rather mysterious component of the natural environment – as such they arouse Man's curiosity and challenge his desire for knowledge, and consequently have had a considerable amount of research effort devoted to them. Furthermore, because of their presence as natural phenomena, they have had a long history of study, which has been intensified in those parts of the world where caves have a direct effect on man's way of life.

Research establishments and university departments around the world have put years of research time into the study of all aspects of caves, mainly their origins, their hydrology and their biology. However the physical agility required to visit many caves means that cave research has been less in the hands of the learned professors than in most other fields of science. Indeed there is a considerable, perhaps unique, overlap between the professional, scientific study of caves and the amateur studies carried out by those who mainly visit caves for the sport. Cave research is dependent on the exploration of caves, and furthermore, there are caves in many parts of the world where the difficulties of exploration make it a full-time task. In these cases a topographic survey is made, probably by the first explorers, and brief structural observations are noted, but there research has to end – the detailed and involved aspects of study, which are the essence of scientific work, are precluded by the problems of accessibility. Fortunately, the combination of adequate knowledge about a few caves with the briefest of knowledge about the rest has been sufficient to provide the scientific world with a reasonable understanding of caves in all their aspects.

The first aspect of caves to receive any attention by the modern scientific world was the very question of how they were formed. Geologists and geographers alike have an interest in caves, in that they are a part of the crust of the earth, with a direct relationship to the surface. Not surprisingly, the first really systematic work on caves and karst came out of fieldwork in the type karst region – around Trieste. At the end of the last century this area was in Austrian territory, so the work came from the Viennese geologists, and in particular Jovan Cvijič. He prepared a doctoral dissertation on the karst region and it was published in 1893 under the title of *Das Karstphänomen.* It was an excellent start, and since then more and more work has led to a better understanding of caves, how they originate and how they relate to surface features.

Nearly all the early geological work on caves was carried out in eastern Europe, in those parts of Austria and Jugoslavia dominated by limestone landscapes. The first centre concerned primarily with caves and karst was also established in the type karst area, now part of Jugoslavia; this was the Karst Institute

Maps and plans of the world's cave systems are made by cavers slowly measuring the passages with waterproof tapes and compasses

at Postojna, which dates back to 1928, sited very close to one of the world's finest cave systems. Since about that time cave research has also developed more in other parts of the world, but still the location of the universities and centres concerned has been closely dictated by the distribution of the cavernous limestones. In England, for example, most work has been carried out by the Geography Department in the University of Bristol, where they have the caves of the Mendip Hills on their doorstep. But in addition much important work has been done outside the universities, by broadly based associations. The National Speleological Society of America and the British Cave Research Association are two fine examples, both of which bridge the gap between the university researchers and the sporting cavers. Both have strong amateur elements, and have produced significant contributions to the science of caves in their own publications.

Next to the basic problems of their origins, caves have attracted most attention to the animals that live in them. Biospeleology – the study of life in caves – has many unique features, mainly due to the lightless and unchanging environment of the underground; most interesting are the various adaptations shown by cave-dwelling animals to their life in the dark. The daily visitors to caves, bats for example, have developed means of 'blind flying', while the extreme is represented by animals who live permanently in the caves, some of whom no longer have any eyes at all, as they would be completely useless. All this provides excellent material for biological research, but unfortunately the caves with the richest variety of life are those in the tropics – mainly remote from population centres and research activities. At the opposite end of the scale the cold caves in the higher latitudes are almost devoid of life. Consequently most of the research has been concentrated in the caves in the areas of Mediterranean climates. The underground laboratories in the caves at Moulis, in the French Pyrénées, are probably the best in the world. Inaugurated in 1947, their most important feature is a series of tanks and pools along the cave stream, where aquatic life can be studied in nearly natural conditions.

Human biologists have also found a special use for caves, in studying man kept in isolation. A cave is an ideal natural isolation chamber where a subject can be kept not only free from all social contact, but also out of touch with the cycles of night and day. He is thus subject to both psychological and physiological stress. Perhaps the most interesting result of this research is the realization that man's 24-hour cycle of sleeping and waking is not entirely built-in, but at least partly dependent on his seeing whether it is day or night. Men who have stayed down caves for periods of three months or more appear to 'lose' up to a third of the number of days in their timeless world below ground.

The more accessible caves in the limestone towers of the southern China karst are still used by farmers as inexpensive barns (right). But prehistoric man in southern France used the local caves in a different way. His beautiful paintings in Lascaux (below) give us insight into his way of life

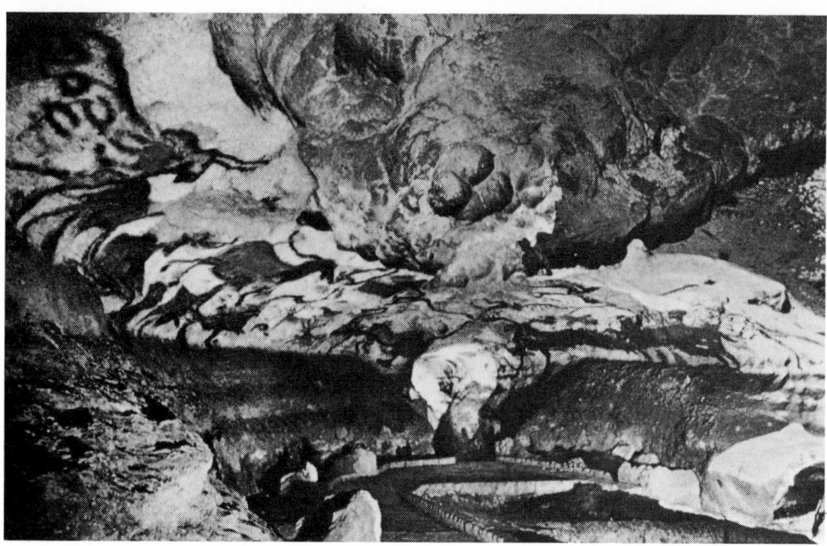

Archaeologists have probably had an interest in caves for even longer than the geologists. Animals and ancient man have long recognized and utilized the shelter offered them by caves. Both have lived and died in caves, and in doing so have left behind clues to their life styles – in some cases, even their skeletal remains. The weatherproof, static environment of the caves has then led to the almost perfect preservation of their entire contents, and caves now provide some of the most valuable material for archaeological research into prehistory. Russell Cave, in Alabama U.S.A., contains a thick sequence of sediments which has yielded a mass of remains; these provide much useful information on the early Americans. The French cave of Lascaux – now almost a household word – is famous the world over for the remarkable paintings on its walls; remarkable not only for their beauty, but because they tell us so much about man's life-style thousands of years ago. Nowhere could these paintings have been preserved so well as in a cave.

Geology, biology, archaeology – they are all studied in caves, and the resulting information is added to the massive human data bank that is known as 'scientific knowledge'. Yet there are also aspects of caves which affect man much more directly. Life, man's included, is dependent on water, and in a karst area all the water is in the caves. So there an understanding of cave hydrology may not be just a subject for research, but a basic necessity. Hydrology is the study of the natural patterns of water flow, and flow patterns underground are the most difficult to study because they are not readily visible. Perhaps the most common problem is to discover the source or sources of water flowing out of caves or springs – so that its quality and reliability can be determined for utilization.

Clearly the most conclusive method of tracing the source of a cave stream is direct exploration. But it is rarely possible to walk up underground rivers for very far, and few subterranean hydrological systems are completely traversible by cavers. Successes have been achieved by this method, but only on quite small, simple systems. The great majority of the world's springs are completely impenetrable, and the hopeful explorer cannot get beyond the limits of daylight. All too often, water flows out through a mass of boulders, or wells up under pressure through the same. In the case of karst springs there must be cave systems behind them, perhaps flooded, but surface weathering has broken down the rock at the very entrance leaving the untouched inner caves virtually inaccessible. Then the karst hydrologist must resort to indirect methods. Water divining has been practised for years, but with dubious results; more recently, refined geophysical techniques have been developed, but these do not always work in natural conditions. The best method of cave water tracing is

to put a marker in the sinkholes and see which spring it comes out of.

Various methods have been used for karst water tracing: perhaps the most successful have depended on the addition of harmless, strongly coloured dyes to the water sinks, and the subsequent detection, either visual or chemical, of the heavily diluted and so extremely weakly coloured waters as they emerge from the springs. Other more modern experiments have used such tracers as radioactive isotopes or the microscopic spores, dyed different colours, of the club-moss Lycopodium. It is by these methods that some quite remarkable underground connections have been established. The Aach Spring, in southern Germany, is the country's largest, and supplies a tributary to the River Rhine. The water emerges from a cavernous limestone and has been proven to originate from a point near Immendingen, seven and a half miles away, where some of the water of the River Danube leaks into its bed. The underground flow therefore crosses the continental divide – it is a characteristic of cave waters that they pay no heed at all to surface topography. The longest karstic underground connection yet known in the world is that of the water sinking from the Beysehir Lake in southern Turkey. This next comes to daylight nearly seventy miles away, at the mouth of the Dudensuyn Cave near the Mediterranean coast.

Probably the most spectacular results of these water tracing experiments are obtained when they go slightly wrong. Not knowing the characteristics of the underground flow makes it very difficult for the hydrologist to know how much dye to put into his sinking stream. A strong coloration at the spring is not really necessary, but too little colour means the test may be a complete failure, so it is only human to err on the high side. The famous French speleologist, Norbert Casteret, did just this when he was trying to trace the true source of the River Garonne high in the Pyrénées Mountains. He eventually achieved success when the fluorescein dye which he put into the Trou du Toro sink came out of the Goueil de Jouéou spring and turned the Garonne bright green for its next thirty miles. The spring at Achabal, in Indian Kashmir, was investigated in 1970 by a team of British speleologists studying Himalayan karst. This spring is regarded as holy, and when asked where the water came from, the local people merely replied 'from the Earth' – a rather unscientific answer. The British team eventually found what they thought was the sinkhole feeding the spring, and to prove their ideas they inserted the red dye rhodamine. Unfortunately the details of the underground hydrology were not as they thought, and two days later the water at Achabal Spring turned almost blood red. As the spring flows directly into a white marble pool in the temple gardens, the effect was embarrassingly impressive!

The Adonis river pouring from its cave mouth in the Lebanon karst (below) is used not only for water supply but also for power in a series of mills. In Morocco a large reservoir is held back in the Chikker Valley just by a tiny dam round the sinkhole which normally absorbs the drainage (far left). Even more useful is the Mas d'Azil Cave in the French Pyrénées which provides a natural tunnel for a main road through the limestone mountain (left)

These dye-tracing experiments have also revealed remarkable variations in the flow-through times for water in cave systems. Dye placed in the Tupper Glacier stream sink, in Canada's Glacier National Park, emerged from Raspberry Spring just over a mile away, in just 53 minutes. In contrast, the dye placed in the Padirac cave river in central France only emerged from the St. George Spring, just over six miles away, after a period of 136 days.

In many parts of the world water from caves is put to good use. Karst spring water is commonly used for drinking supplies – indeed in some areas it is the only available supply. The Zagros Mountains of southern Iran have almost a desert climate, yet they consist partly of limestone. Consequently villages have grown up around many of the springs, as away from them water is only available by digging wells. Most of the water emerges from piles of boulders, though in a few cases it is taken from open caves. The sources of the springwater are nearly all unknown; this scarcely matters in these cases – the experiences of time have shown the mountain peoples which springs are reliable and do not dry up in summer. But any larger concern utilizing cave waters must have a good knowledge of the catchment areas which feed the springs to be able to evaluate their usefulness. The fast flow through time, and negligible reservoir capacity, of most cave streams makes them prone to high flood peaks and very low summer flows. Hydro-electric undertakings which use cave rivers, such as that from the Bournillon Cave in France, have to allow for such high flow variations. Another sort of problem was met by a Spanish hydro-electric power company working in the Pyrénées; they had planned to capture the waters draining from the Maladetta massif before they disappeared into the Trou du Toro sinkhole; they would then release the water lower down the same valley into the Rio Esera (which flows to the Mediterranean). However Casteret's spectacular dye test, mentioned above, showed that the Trou du Toro water flowed underground beneath the mountains to the River Garonne, and in fact provided the main summer flow for the river, where it was utilized in the high valleys just over the frontier in France. Any diversion of the water would therefore leave these French valleys almost dry, and the Spanish authorities had to abandon their power scheme.

Surface rivers are of course known for their vulnerability to pollution. In contrast, groundwater pumped out of wells or tapped from springs is generally regarded as pure. This is commonly true if, for example, the water has flowed through a sandstone, for then it will have probably spent years underground, and in working its way through the network of pore spaces will have been effectively filtered and purified. The same does not apply to water coming from a limestone spring – its rapid flow

In Missouri, USA the underground flow from the Dora Sinkhole rubbish dump (above) to Hodgson Mill Spring (above right) was proven by tracing with dyed lycopodium spores –

which are ten times the size of typhoid bacilli. A holy temple has been built around the deep spring below at Verinag in the Vale of Kashmir (below)

through open caves, really underground rivers, will have had little or no filtration or purification effect. Therein lies an often unrecognized danger inherent in karst spring water.

The Achabal spring in Kashmir is used directly by the local people for drinking water, for they would never dream of drinking from the adjacent river whose water is foul and muddy from its passage through endless rice paddies and other villages. Yet the spring water has its source in leakages in the limestone bed of that same river about nine miles upstream. The two-day underground journey of that water, through what must be quite sizeable undiscovered caves, has allowed some of the mud to settle from it, so the spring water is normally quite clear, but it can have resulted in very little other purification.

Such problems of cave transmitted pollution are not helped by the widely held opinion that anything which goes down a sinkhole is never seen again. Everything comes back somewhere! Nedge Hill Sink is a small cave on the Mendip Hills of England. In 1967, a Bristol cleansing firm dumped 50,000 gallons of oil and cyanide into it. There are numerous springs in this part of the Mendips and unfortunately it is not known which is fed by the Nedge Hill Sink. Somewhere this water, oil and cyanide mixture must have been fed out onto the surface of this pleasant and heavily farmed countryside.

One of the most disturbing cases of cave transmitted pollution comes from Ireland. The Ballymacelligott Caves, near Tralee, appear on the surface as a series of short sections of collapsed cave and surface valleys spread out for just under a mile down a hillside. Altogether, between the top sink and the bottom rising,

Safe pathways and good lighting make it easy to see just what is there in a show cave, such as America's Mammoth Cave (below). But in an undeveloped cave everything is difficult: mapping a Venezuelan river cave (right) involved taking survey notes while swimming. The photograph of the Main Chamber in England's Gaping Gill Cave (far right) was made by leaving the camera shutter open for a minute, to pick up the daylight on the waterfall from the surface, and also firing five large flashbulbs

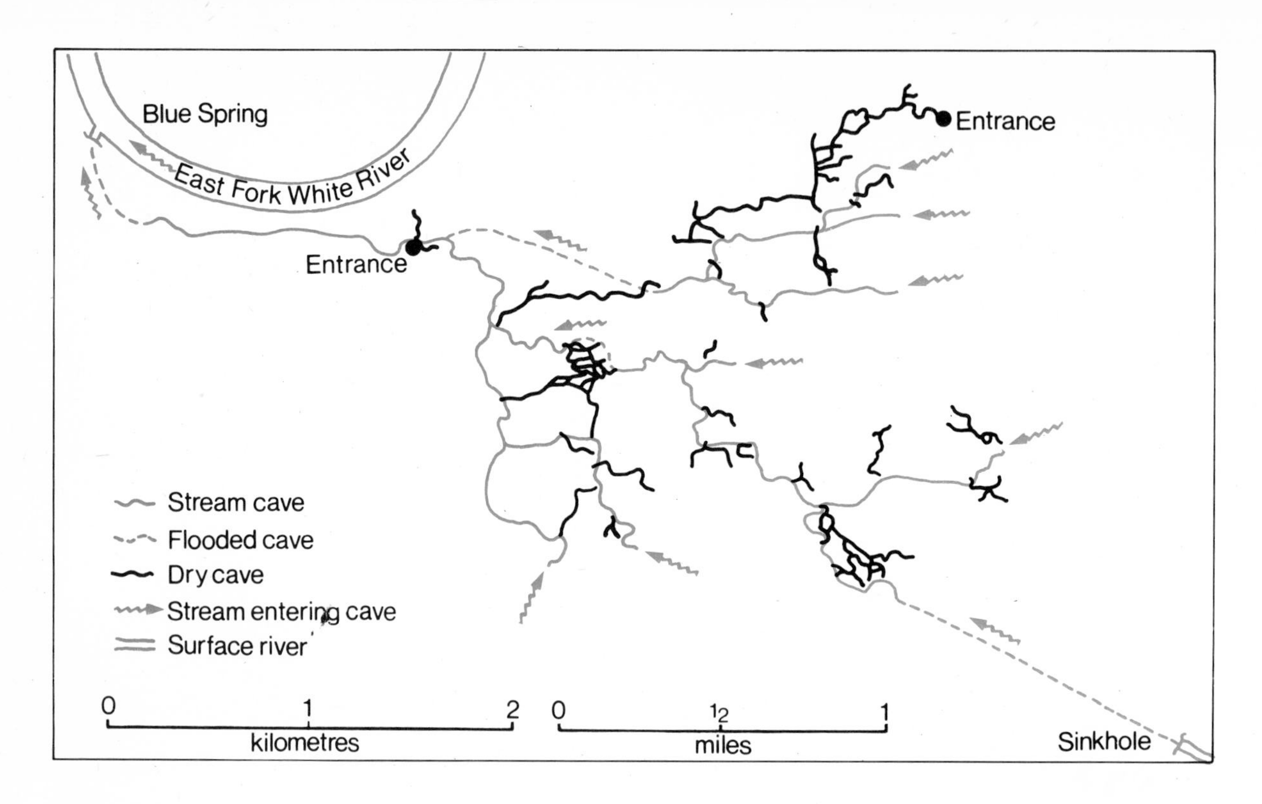

Blue Spring Cave, Indiana, contains a converging series of separate streams together with networks of abandoned passages

there are seven of these surface openings. It was only realized that all nine openings were in the same stream when a party of Bristol cavers examined the area in 1964 and explored some long sections of cave passage between the various openings. Then the pollution hazards were appreciated for the first time. At the third entrance from the top the stream was used by cattle for drinking, and other purposes. At the sixth opening down, the stream was used for a house supply, and gathered the lavatory sewer which fed directly into it. The seventh opening was again used by cattle. At the bottom rising the stream flowed out of fissures in the limestones and was used to supply a farm and a pair of cottages. This was after the water had already acted as a lavatory sewer and had been 'used' by cattle – without any natural filtration as it flowed through the open caves!

Civil engineers have been concerned with caves for as long as their profession has existed, for man made many of his first homes in caves. This early adaptation of caves barely involved civil engineering as we know it today. Fortification of cave homes probably reached its peak when the Predjama Castle was built in the entrance of the cave of that name in northern Jugoslavia – at least the defence of its back doors presented little problem. Caves have also been of military significance, particularly for example in the use of the thousands of caves of Vietnam,

One of many houses in the town of Bank, South Africa, which has collapsed due to catastrophic subsidence of the soft sediments overlying highly cavernous limestone

in the guerilla warfare in that country. The more scientific approaches of civil engineers to caves really originated from the first major engineering feat in cavernous country – the construction of the Vienna-Trieste railway in 1857, right through the classical karst. There the problem was not to utilize the caves, but to avoid them.

In cave country engineers have always had extra problems, due to the possibility of construction plant disappearing down holes when thin cave roofs collapse. The problem is all the more acute because of the notorious uncertainty of the precise location of the cavities. However such collapses are rarely of the dramatic type where vehicles disappear into gaping holes; normally there is just a sharp subsidence, and any constructions involved merely end up on a great heap of rubble a few yards lower than they originally were. One of the few cases known of such collapses opening directly into a passable cave occurred in the winter of 1946, high on the limestone fells of the English Pennines – well away from any building. There the floor of what was a shallow closed depression just dropped away, probably due to a heavy rainfall, leaving a forty-five foot deep shaft and wide open cave entrances in each side. These led directly to a cave system more than half a mile long and nearly 500 feet deep – now known as Notts Pot.

Changes in hydrology are usually the cause of roof collapses, for changes in water conditions will result in variation in the strength of any loose sediment covering the limestone – and it is usually the collapse of such sediments bridging cavities, rather than the collapse of limestone cave roofs themselves, which causes the problems. The most extensive series of collapses recorded has been in the Carletonville area of the Rand, South Africa, and these have been essentially triggered by removal of water from the sediments on top of a thick series of limestones, due to pumping in connection with deep mining in the area. Over a hundred sinkholes have developed in this way; the largest is more than 300 feet across and over 130 feet deep. Unfortunately these lie in an urban area and there has consequently been considerable damage to buildings and the loss of over thirty lives. The U.S.A. has also had its share of sinkhole collapses, again many related to hydrological changes, mainly in the states of Florida and Alabama, both with huge areas floored by limestone. Perhaps the most spectacular collapse occurred in Lexington, Kentucky, when in 1965 a truck dropped thirteen feet into a cave – while it was pouring concrete for the foundations of a house!

Not surprisingly miners working through cavernous limestone have frequently had very serious water problems. One of the greatest inrushes of water ever recorded was in the Morococha copper mine in Peru; a heading intersected a flooded cave from which the water poured at a rate of 700 gallons per second, though it eventually slackened off as the cave system drained dry. Lead miners working in Halkyn Mountain, North Wales, had perpetual problems with water inrushes and flooding from the many caves in the limestone. Eventually mining could only continue when the whole mountain was drained out by a mined tunnel excavated along its length and out to the coast at Holywell. This tunnel serves as a convenient transport route out of the mines, but half its floor width has to be devoted to a six foot deep and wide channel which carries a ceaseless flood of water out of the caves.

To compensate for all the trouble that caves cause to engineers there are a few cases where they have been positively helpful. The road from St. Girons to Pamiers in the French Pyrénées was built down a river valley which suddenly terminated in a precipitous limestone cliff, where the river passed through the Mas d'Azil cave and out of the other side of the ridge. With the only alternative a steep climb, the engineers did the obvious and ran the road through the cave, for it averages sixty-five feet in width and thirty odd feet in height, and descends only gently. Little engineering work was needed and the road now runs for the full 500 yards or so through the cave, on a ledge just above the rocky riverbed. Midway through, a chamber opens out and encloses a small car park – for a commercial show-cave has been made out of some decorated side passages; but the main river passage provides probably the best cave in the world which is lit up and accessible to tourists, for free.

Even more remarkable is Natural Tunnel, a cave, 300 yards or more long, through Powell Mountain in Virginia, U.S.A. This was only discovered in 1880, by J. H. McCue, an engineer surveying a route for a railway. It was then nearly full with sediment and driftwood, but when this was cleared out the 100 foot diameter of the cave was more than adequate to take the Bristol-Appalachia railway line without any further engineering works. Construction of a railway in the Harz Mountains of Germany revealed a cave, the Himmelreichhöhle, when a tunnel was driven through a small hill near the village of Walkenreid. The cave consists of only a single chamber with no passable connections to the outside, and the railway goes right along its length of over 300 feet. However this cave is formed in gypsum, a rock much weaker than limestone, and the roof was so prone to breakdown that the railway could not be left unprotected in the cavern. So a concrete tube was placed around it, across the floor of the cave. An access door has been left through this, so that visitors to the cave now gain the unique experience of walking along a railway tunnel, on the outside!

In common with engineers, miners normally regard caves as an inconvenience. Yet there are cases where materials have been extracted from the caves themselves. Guano is a phosphate- and nitrate-rich sediment formed almost entirely from the droppings from birds or bats. One of the environments where it most rapidly accumulates is on the floors of large cave chambers

In Kimalia Cave in Kenya the bat guano is mined, packed into sacks and hauled out through a hole in the roof

which are used by thousands or millions of bats for their daytime sleep. Guano is also of commercial value, the phosphate for fertilizer and the nitrate for the production of gunpowder. In the U.S.A. two of the largest known caves have had their guano deposits extensively mined. Mammoth Cave was worked during the war of 1812 for its nitrate component, and Carlsbad Caverns had 100,000 tons of guano shipped out for fertilizer in the first few decades of this century. Also worked for guano is the Niah Great Cave in Borneo. But like many other caves in the East this cavern is better known for the edible birds nests which are found in it. It is the salangane birds which construct these nests, from which the popular delicacy of 'bird's nest soup' is made. The nests are attached to the walls and ceiling of the caves; in Niah they are over 150 feet from the floor in places. Consequently it is a death-defying occupation for the collectors, who gather the nests with bamboo poles from the tops of frail structures also built of bamboo. One variety of 'cave-mining' has now ceased, and will never be allowed to restart while man cares for the environment. Early in the last century, Wookey Hole in the English Mendips was invaded by a team of musketeers who were directed to shoot down the stalactites off the roofs of the great chambers. The broken formations were then gathered up and taken to London, where the end result of this incredible desecration was merely the decoration of a suburban ornamental garden! It is a refreshing contrast that a visitor to the show cave now open in Wookey was recently heavily fined for breaking off one of the well-known stalagmites 'for a souvenir'.

Caves have been most useful to the commercial world in their development as show caves, suitable for visits by large numbers of non-caving tourists. Massive underground chambers and large numbers of stalagmites and stalactites provide a natural spectacle of the first order. It is of course a feature of show-caves that they are not typical of caves in general, for there is no way in which a coachload of tourists can be conducted along a wet crawl passage half a yard high. But this does not really matter, for show-caves only attempt to display some aspects of the underground world, and this they do very well; indeed, visiting a careful selection of the world's show-caves can give a traveller a very clear impression of most aspects of caves. Large chambers and beautiful formations are the stock-in-trade of most show-caves, but some of the world's finest ice formations are to be seen in the Dachstein and Eisriesenwelt show-caves in Austria, and the Piuka Jama show-cave in Jugoslavia gives the visitor a very close view of a large underground river. Underground boat rides are such an attractive subject that the show-caves of the world probably have a disproportionately large share of the cave lakes that are known.

The water has been largely drained out of the Ingleborough show-cave in the English Pennines so that a concrete path could be laid for the convenience of the parties of visitors. The original water level in the cave is marked by the bands on the rocks and stalactites five feet above the present floor

There is, not surprisingly, considerable variation in the ways in which caves have been developed for tourist purposes. Some caves are now visible in well preserved natural conditions, while others have been so changed 'in the interests of commerce' that they are barely recognizable. The larger caves have a certain advantage here, in that paths can be laid in them so that they barely intrude on the natural conditions. Some of the great caves in the American National Parks, Mammoth Cave for example, are characteristic of the well developed show-cave, and they have clearly gained from being in the Parks system and therefore free from competitive commercial interests. The same caves are also notable for the cleverly concealed lighting and the burial of the necessary power cables. At the opposite end of the scale are some truly dreadful show-caves which are best left unnamed. Typically they seem to have large proportions of mined passages, connecting relatively small natural cavities, dirty and muddy calcite formations, clumsily laid footpaths sur-

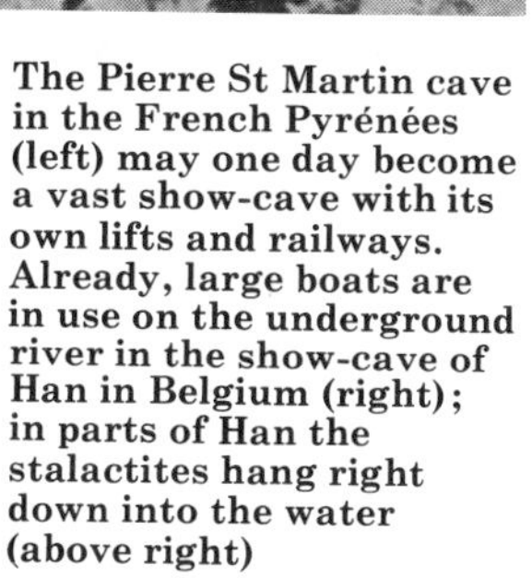

The Pierre St Martin cave in the French Pyrénées (left) may one day become a vast show-cave with its own lifts and railways. Already, large boats are in use on the underground river in the show-cave of Han in Belgium (right); in parts of Han the stalactites hang right down into the water (above right)

rounded by vast areas of fencing and netting, and naked light bulbs attached to the walls at convenient intervals. Lakes in such caves are commonly held up by concrete and the most natural thing about them is the mud. Even worse, and not uncommon in these caves, is the planting of stalagmites. There are a number of cases where rather inaccessible caves have been stripped of their formations, which are then taken to enhance the ledges and grottoes of a show-cave. Cruder still is the building of concrete stalagmites with the hope that they will look natural when coated with a thin layer of calcite from a convenient drip from the roof. Unfortunately this does not always work; the concrete may be eroded instead, and there is a locality in a well-known English show-cave where a close look from an awkward angle reveals the steel framework inside a 'stalagmite'. The patter from cave guides also varies greatly in quality. Some do give most interesting and scientifically correct stories concerning the caves and their history; others merely point out odd shaped stalagmites with ridiculous names–'Mother and Child' is a favourite – and give vastly exaggerated dimensions for the bits of the cave which the tourist cannot see – experience has shown that lengths are normally multiplied by a factor of somewhere between five and ten!

Floodlighting for tourist purposes has been invaluable in some caves, where the passage and chamber sizes are so great that they just would not be visible with a normal caver's lamp. In this context the Carlsbad Caverns of New Mexico, U.S.A. must be mentioned. Run as a show cave by the National Park Service, they provide a marvellous experience. Entering the cave the visitor walks, all downhill, for about an hour, through huge galleries and down immense boulder piles, to a depth of over 800 feet, where a rest can be enjoyed in a restaurant in a side chamber. The tour continues with over an hour's walk round the perimeter path of the Big Room – one of the largest chambers in the world – which is festooned with every type of stalactite and stalagmite. Fortunately for some, the way back to the surface is in an elevator from a corner of the restaurant chamber. Europe's equivalent to Carlsbad in terms of 'large show-caves' must be the relatively little known Skocjanske Jama cave in northern Jugoslavia. There the visitor starts his tour with an easy walk along a large gently descending passage decorated with lines of massive stalagmites and columns. This is impressive enough, but the passage then opens high in the wall of the main river passage. This is an almost unbelievable size – more than 300 feet high and 150 feet wide. The tourist path turns along a ledge, crosses a bridge to the far wall, and continues upstream on ledges and terraces halfway up the side of this massive underground canyon – eventually to daylight at the main river entrance. The huge floodlights, which were installed to illuminate the magnificent views from the path, revealed features of the cave which had never previously been seen by the cavers and engineers, who had worked just with small personal lamps.

Not far from Skocjan are the well-known Postojna caves. These were first developed for tourist visits in 1818, and in 1872 gained the distinction of a railway laid from the entrance for over a mile into the heart of the cave system. To this day the trains still carry thousands of visitors through the stalagmite-laden galleries, to where the footpaths lead through the central complex of large chambers.

This railway is already a reality, but plans for other show caves have been even more spectacular. The huge Eisriesenwelt cave system in Austria has its first half-mile or so of passages liberally decorated with ice formations, and these are open as a very fine show-cave. But in the 1920s Herman Gruber dreamed of opening the far reaches of the cave to tourists. He planned to

take visitors for another half-mile along the vast Midgard tunnel and then by lift 800 feet up a natural shaft to the plateau above. With this in mind he started to build a railway line through the Midgard, and he did nearly all the work on his own, frequently staying for two weeks at a time in the depths of the cave. He'd built over 200 yards of embankment winding through the boulder piles, when he found there was no usable connection to the plateau above, and abandoned the whole project. Within the last few years magnificently grand plans have been proposed to turn the world's deepest cave, the Pierre St. Martin system in the Pyrénées, into a show-cave. These involve a series of lifts down the 1000 feet deep Lepineux shaft, then a route, possibly via underground aerial ropeway, through the series of massive chambers, and a monorail ride through a bored tunnel back out to daylight in the side of the valley, over 2300 feet below the top entrance. A fabulous scheme for a fabulous cave.

The stalactites and stalagmites loom over the orchestra which regularly plays in the huge main chamber of the St Michael's Cave deep in the Rock of Gibraltar

5 Formation of Caves

Caves owe their existence to the solution of limestone in water. The erosion of limestone by perfectly natural methods, principally by solution in water, takes place not only on the surface, but within the rock itself – where it is eaten away along fractures and lines of weakness originally created by earth movements. The underground results of this process are the caves as we know them. However it may be easily seen, in almost any cave, that many other processes are taking place, all contributing to the enlargement of the passages. Underground rivers, particularly in times of flood, carry heavy loads of sand and mud; in moving these along, the rivers effectively abrade their bedrock, carving passage floors deeper into the limestone. Blocks of limestone are also undercut, and fall from the cave ceilings; they may fall into the cave rivers and be broken up, some parts to be carried away in solution and other parts to be transported out of the caves as limestone pebbles or boulders. Essentially most of the processes present in a surface river channel also take place in an active cave passage – but this analogy only applies to large cave passages. In the microscopically narrow fractures from which nearly all caves must initially develop, sediment transport, collapse and abrasion cannot take place. Therein lies the importance of solution, for while solution is just one of many processes active in a large cave passage, it is the only process taking place in these initial stages.

Limestone is by far the most common rock which is naturally soluble in water. But it is not the only one. Indeed, caves may be formed in rocks which are not soluble at all – in lava and sandstone for example – but in these cases completely different processes are involved. Of the soluble rocks other than limestone, gypsum and salt are the most important. Neither rock is very common, and both are soft and highly soluble in water – so much so in the case of salt that the entire rock tends to be eroded away or deformed even more rapidly than caves are developed. Consequently caves in gypsum and salt are not as common as in limestone. Dolomite is another rock containing caves, but it consists only of the double carbonate of calcium and magnesium – which is very similar to the calcium carbonate of which limestone is formed; the differences are so slight that cave formation in dolomite and limestone may be regarded as essentially the same process.

Rock salt consists of sodium chloride; 26 grams of it may be dissolved in a litre of water – or, expressed another way, its solubility in water is 26,000 parts per million (ppm). Gypsum is hydrated calcium sulphate and its solubility is about 2000 ppm. From these high solubilities it is easy to visualize how caves may be formed in either rock by percolating rainwater. Yet limestone, consisting of calcium carbonate, has a solubility in pure

Valley glaciation in the English Pennines has been largely responsible for the present variety of passage types in the cave systems

water of less than 15 ppm. Such a limit would impose an exceptionally long time requirement on the formation of large cave systems, but fortunately the solubility of limestone in water is strongly influenced by the amount of carbon dioxide in the water. Some naturally occurring waters may dissolve over 400 ppm of limestone, and figures such as that make cave formation in limestone a more reasonable process. This dependence of limestone solution on an external factor, the carbon dioxide availability, is unlike the straight-forward solution of salt and gypsum, and contributes to the great variety of caves and cave features found in limestones.

The precise chemistry of the limestone solution reaction is extremely complicated, but the significant parts of the process are fortunately quite simple: calcium carbonate is only negligibly soluble in pure water, but calcium bicarbonate is much more soluble; and it is the formation of the bicarbonate, by reaction with the carbon dioxide and water, which has the overall effect of limestone solution, and at the same time demonstrates the importance of the carbon dioxide. Rainwater absorbs carbon dioxide from the atmosphere, but only enough to give it a solution capacity of about 50 ppm of limestone. The greatest natural concentrations of carbon dioxide occur in the soils, and it is water that has percolated through an organically rich soil which may be capable of dissolving around 400 ppm of limestone. The carbon dioxide content of the soil is controlled by the level of organic activity, and clearly this is related to climate. Limestone solution is much more rapid in a tropical forest than in a thinly vegetated arctic tundra; caves have similar forms in each climatic zone, but are developed at very different rates.

More important than this in terms of cave development, are the variations between carbon dioxide levels, and therefore limestone solubility limits, of different waters in one area. Soil water, saturated with a high content of calcium carbonate, may pass into contact with normal air – perhaps in a cave; carbon dioxide then diffuses from the water to the air, to maintain equilibrium. The result is that the water is then over-saturated with calcium carbonate, so some is deposited; therein lies the principle mechanism behind the formation of calcite stalactites and stalagmites (see Chapter 6). In addition soil water contains very much more carbon dioxide than river or stream water when they enter the limestone. But it is obviously possible for both types of water to become saturated very rapidly as soon as they come into contact with the rock. How then are caves formed deep below the surface where the water in them has already been saturated? The answer lies in the difference between the waters, for when two waters, containing different amounts of carbonate in solution, mix, then even if they were both saturated the newly formed

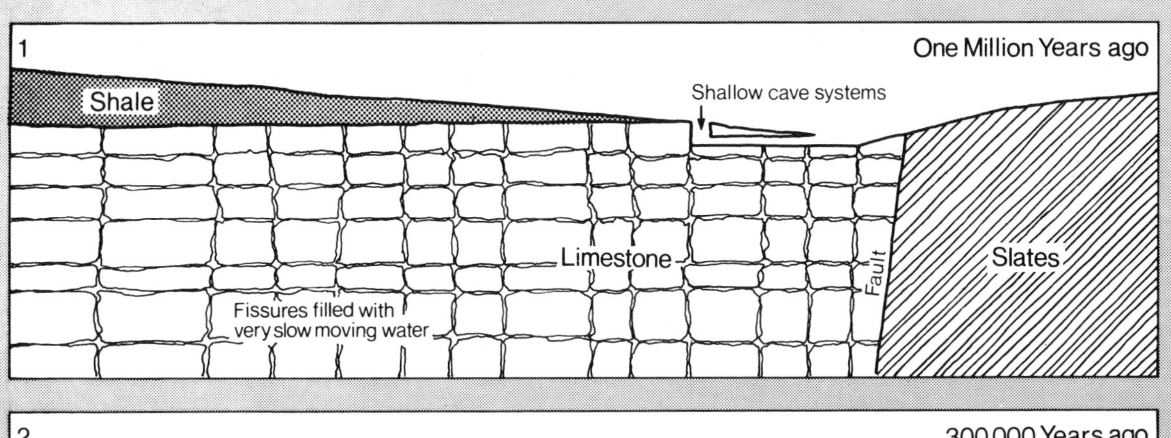
1
One Million Years ago
Shale
Shallow cave systems
Limestone
Fault
Slates
Fissures filled with
very slow moving water

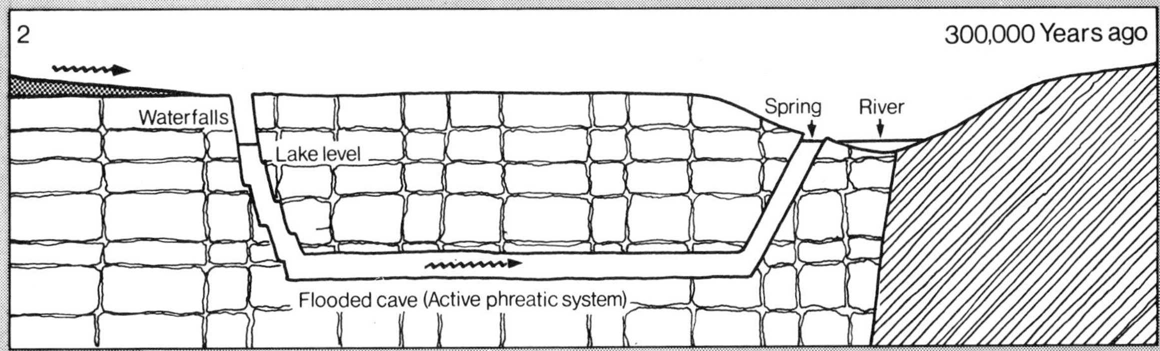
2
300,000 Years ago
Waterfalls
Spring
River
Lake level
Flooded cave (Active phreatic system)

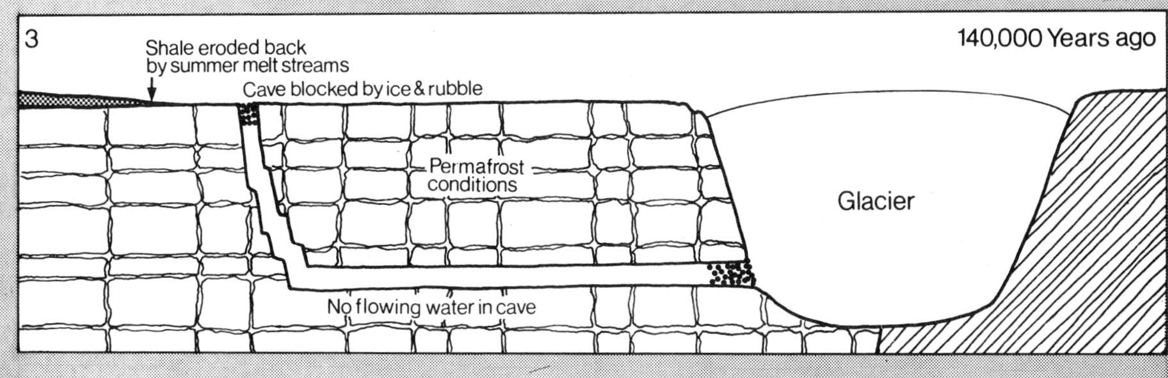
3
140,000 Years ago
Shale eroded back
by summer melt streams
Cave blocked by ice & rubble
Permafrost
conditions
Glacier
No flowing water in cave

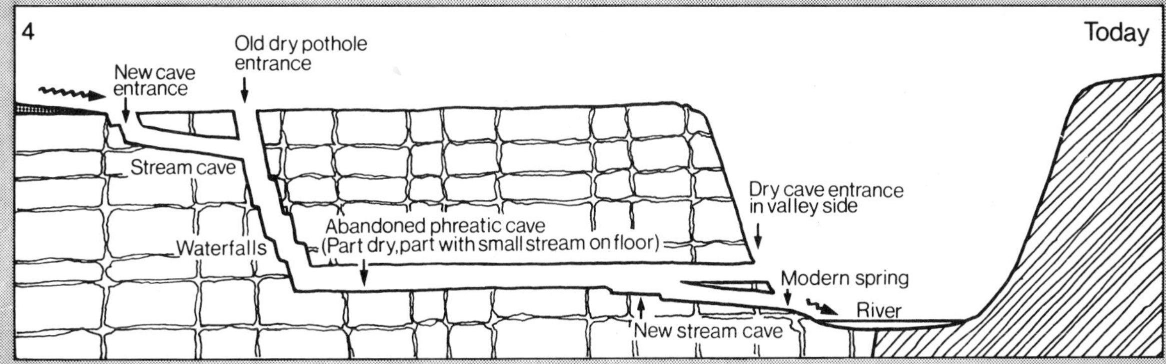
4
Today
Old dry pothole
entrance
New cave
entrance
Stream cave
Dry cave entrance
in valley side
Abandoned phreatic cave
(Part dry, part with small stream on floor)
Waterfalls
Modern spring
River
New stream cave

mixture is capable of further solution of the limestone. This important concept of 'mixed water corrosion' was first recognized and described by the famous Swiss speleologist Dr. Alfred Bogli in 1964, and it makes a major contribution to the understanding of the mechanisms of cave formation.

Water can only form caves when it has a ready-made access route into the limestone itself – even if it is only a microscopic crack. Without such initial fractures, water can only bore its way into limestone for a very short distance – a few feet at most. So the pattern of these fractures has a very marked influence on cave development. Indeed caves can only form where there is such a network of fractures, perhaps spaced every few yards or so in a relatively strong, non-porous limestone. The rather weaker, more porous varieties of limestone (the English Chalk is the best example), tend to have fewer fractures as they have not broken so easily under the stresses developed by earth movements. Yet these rocks still absorb vast quantities of water and result in a streamless landscape. Their groundwater flows through the network of pore spaces, in effect making micro-cave

Water flowing through a ruler-straight joint in the limestone has opened a narrow passage in Lancaster Hole (far left) in England. Walls have been carved into delicate scallops by eddies in the turbulent water (left). Wider, more rounded scallops have been left on the bedding plane roof of an ancient stream passage in Italy's Monte Cucco Cave (above)

systems, and the water is not significantly concentrated in the larger joints where it could form man-size cave passages. Consequently the English Chalk and most other similar porous limestones are almost devoid of caves.

Caves are formed instead in the massive, strong, non-porous varieties of limestone. Then no water can pass through the rock itself and all the flow is concentrated into the fractures where proportionately large cave passages are carved out. Millions of years before cave development ever started in a bed of limestone, the rock was subjected to stress, due to great earth movements. Under most conditions limestone is a fairly brittle rock and the result of this stress would normally be a network of joints – simple breaks in the rock, similar to those formed in a broken window before it falls out of its frame. In some cases more intense earth movements have caused movement along these fractures when one block of rock slides past its neighbour. The resultant fracture is known as a fault and is frequently recognizable by a zone of broken rock, crumbled and smashed as the rocks move past each other. Faults tend to act as more open channelways for the initial flows of groundwater, but they are rarer than joints.

Another very special type of fracture is the bedding plane. Limestone is originally formed as a sediment on the floor of a shallow sea, and like all sequences of sediments it is not completely uniform from top to bottom. Different layers, or beds, may have slightly different compositions or textures, and these differences tend to concentrate stresses along the intervening bedding planes, commonly resulting in fractures along them. Even more important structures in limestone are thin beds of shale – representing layers of mud or volcanic ash laid down as breaks during the sequence of carbonate deposition. These are commonly only a few inches thick – forming less than one per cent of the total limestone sequence – but being watertight they have a great influence on the initial flows of water in the fracture networks.

It is all these features – joints, faults, bedding planes, shale beds and in addition the different solution rates of different limestone beds – which influence the patterns of water flow, and hence cave development. As soon as any part of a limestone mass is exposed to the surface, water begins to circulate within it. Rainwater enters from the surface down the joints into the limestone. It flows along bedding planes and down further joints. It collects on the watertight shale beds and flows along on top of them till a large enough joint or fault opens a path through the shale to further depths. Under pressure the water may also flow uphill, utilizing the same sets of fractures and routeways through the limestone.

In a typically fissured limestone the result of this first estab-

Cave development commonly starts in the bedding planes in the limestone. A bedding plane exposed by rock fall in Mammoth Cave, USA, (right) contains a network of little tubes which were the first channelways to form through the rock. A bedding plane also first guided the water through the Long Churn Cave in England (above); the stream has since cut a trench in the floor but left the flat roof untouched

Cave formation

The formation of caves means moving water. The Little Neath River carves out its own cave (centre left) in the black limestone of South Wales. P8 Cave in the English Peak District (far left) is formed in a great sloping fault in the rock. A vertical fracture has also been opened out in the Epos Chasm in Greece (below); the turbulent spray of the huge waterfall in the Chasm each winter has enlarged the initial fracture into a cylindrical shaft. The solutional attack on the limestone is aided by undercutting and collapse, such as in Italy's Piaggia Bella Cave (left)

Sequential development of a decorated cave

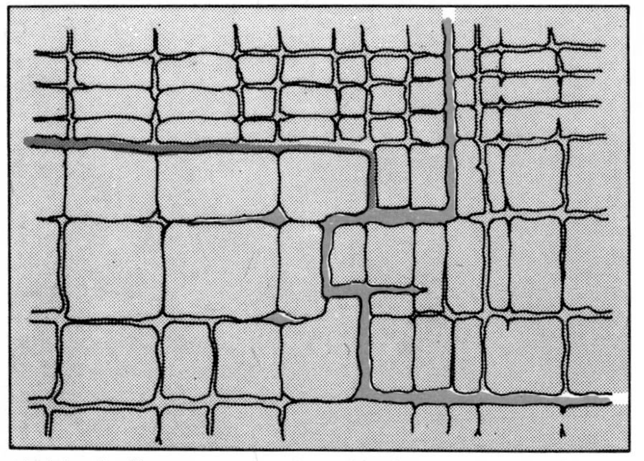

1. Water in fissures – moving slowly

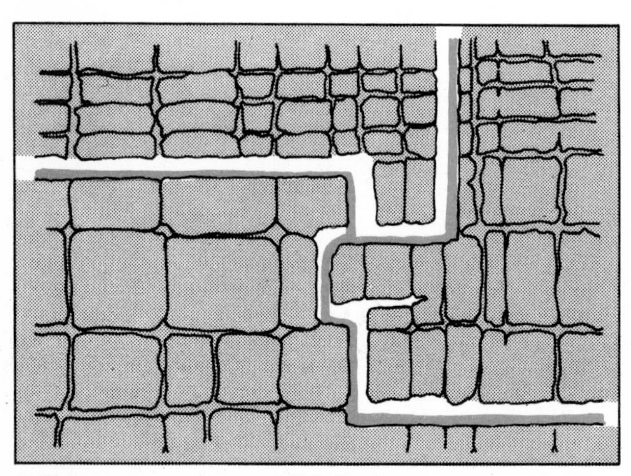

2. Stream caves are established

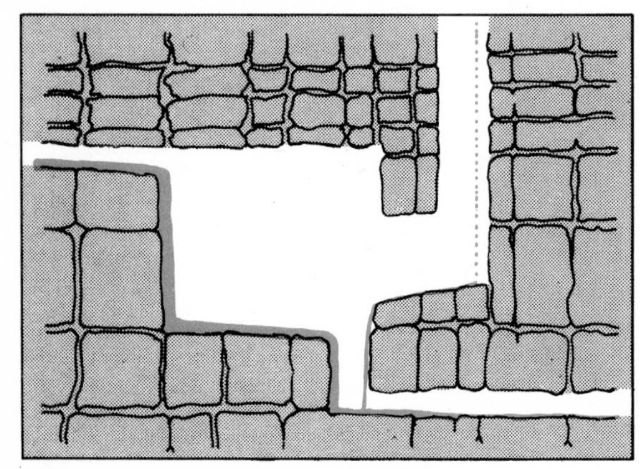

3. Stream erosion cuts out chamber

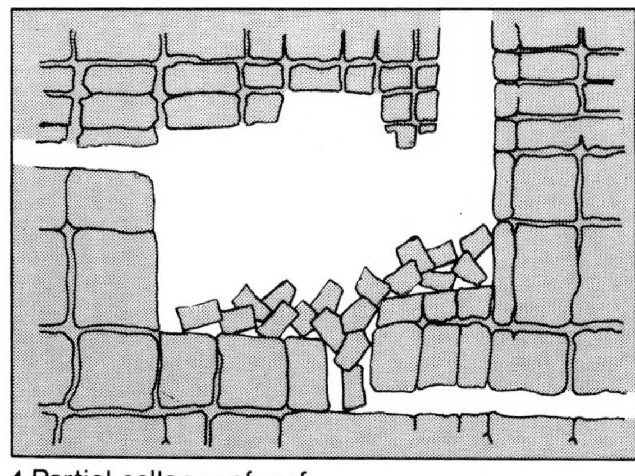

4. Partial collapse of roof

5. Solution of collapse debris and more floor

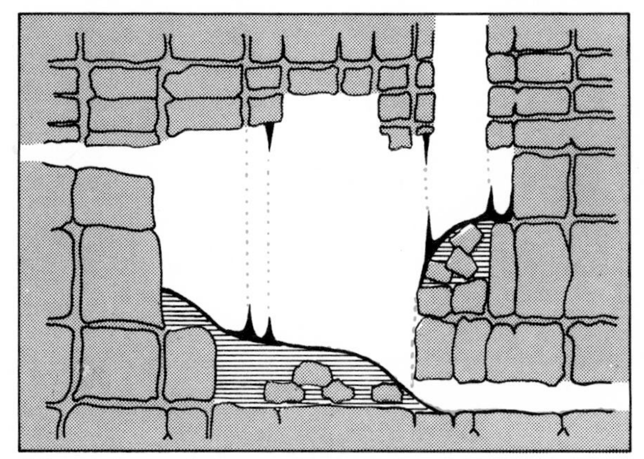

6. Deposition of clay sediment followed by stalagmites

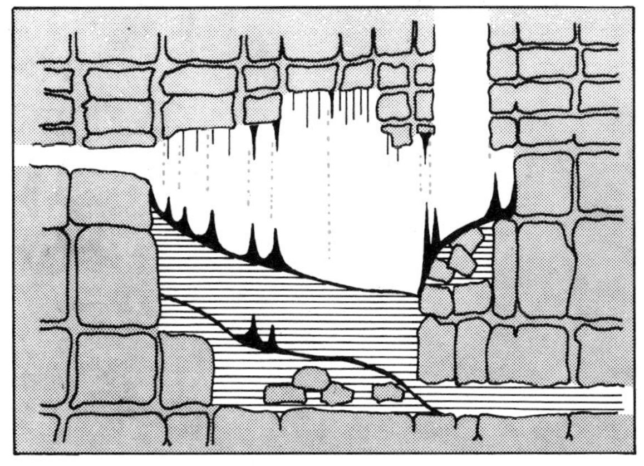

7. Deposition of a second phase of clay sediment followed by more stalagmite

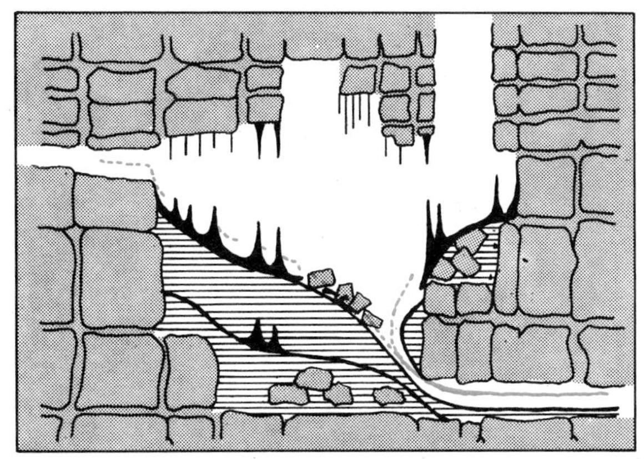

8. Erosion of some of the clay sediment and stalagmite floor, and further collapse

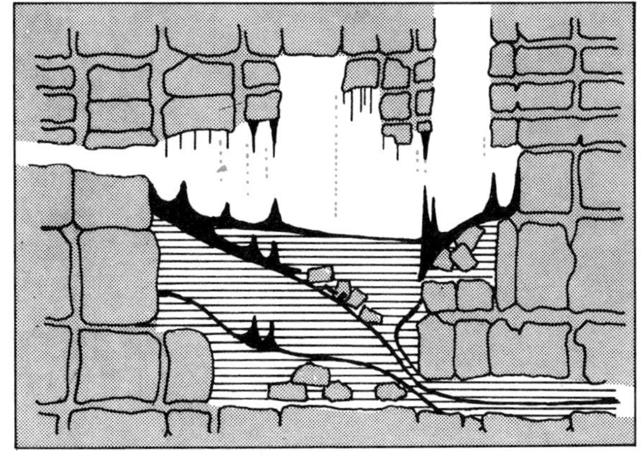

9. Deposition of a third phase of clay sediment and stalagmite

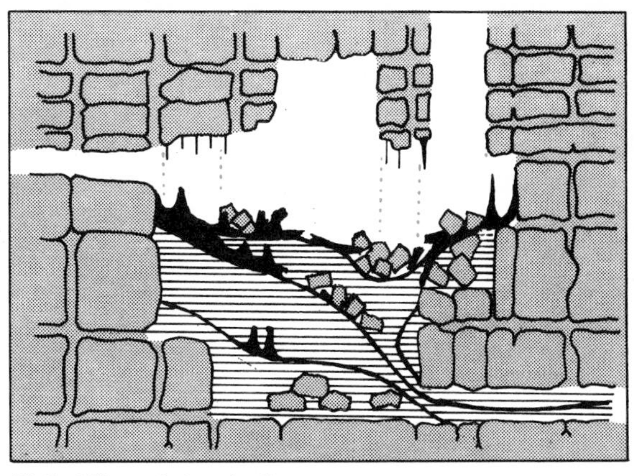

10. Settling of clay sediment, some collapse of stalagmite and further roof collapse

11. Modern phase of calcite deposition

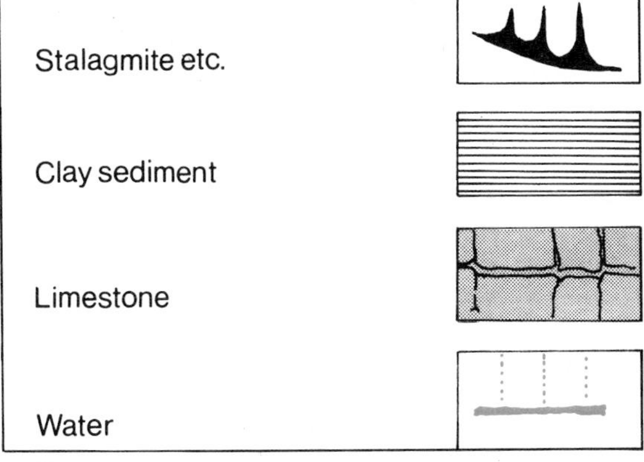

Key

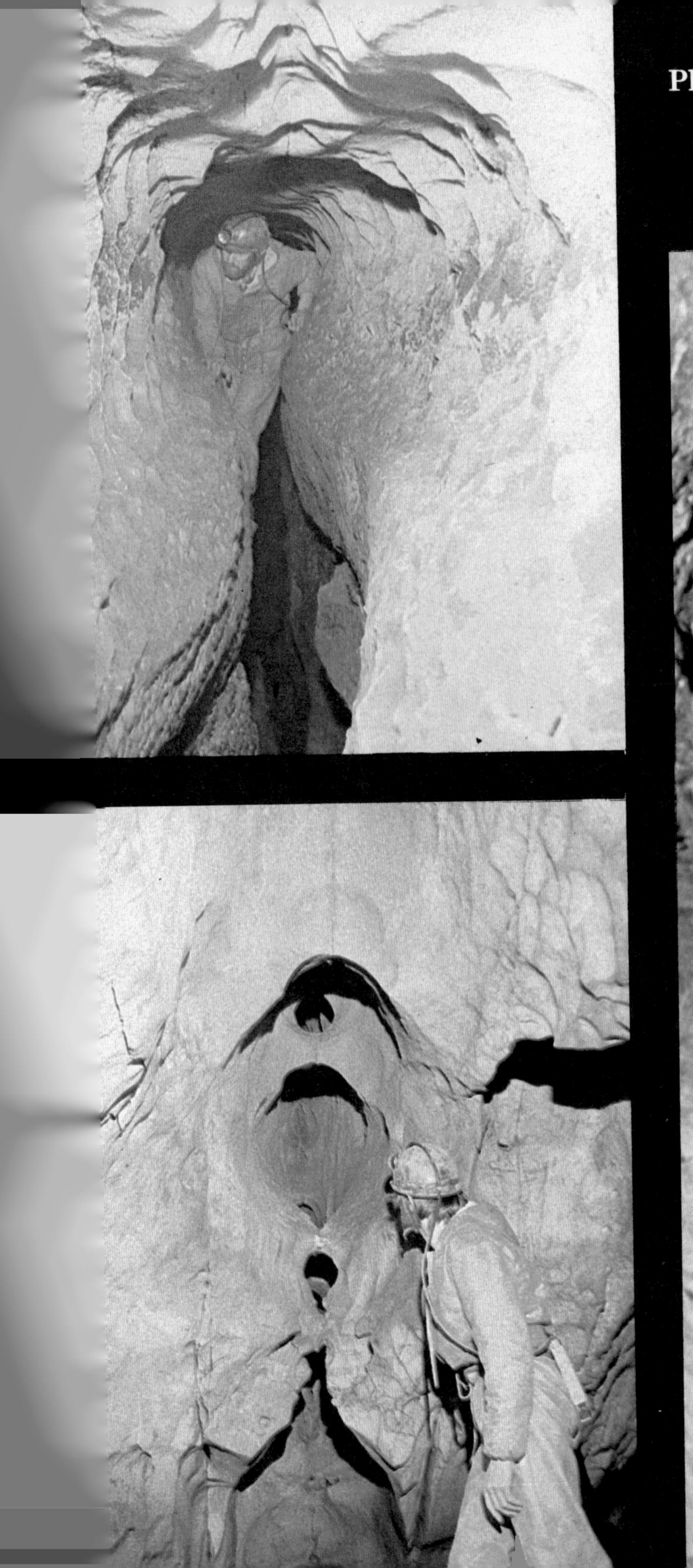

Phreatic tubes

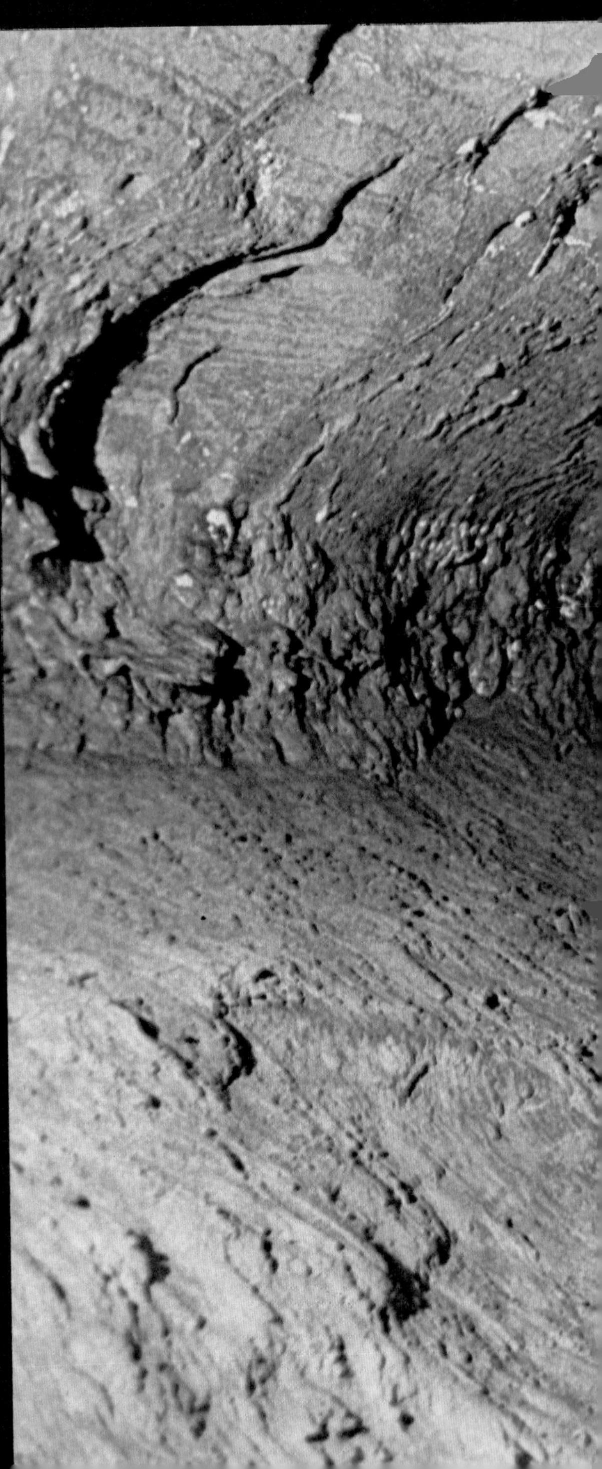

Phreatic passages formed by water under pressure tend to exhibit rounded sculpture of the rock. The Subway in Canada's Castleguard Cave (below) was bored along a fracture in the rock by a river completely filling the passage. The keyhole shape of a passage in England's Gaping Gill cave (left) shows that it too was a phreatic tube before the slot was cut in its floor. In the same cave, rounded pockets were etched out along roof joints (below left) when the passage was full of water.

lished flow of water is an immensely complex three dimensional network of water-filled openings. The widths of these initial fissures is measured in microns – thousandths of millimetres. The forces of friction and surface tension are so high in these narrow openings that the water moves extremely slowly. Gravitational flow is established, with water entering the limestone on the higher ground and pouring out of springs which mostly form at the lowest points where the limestone meets the surface. The routes taken between these two ends may be extremely circuitous, as dictated by the details of geological structure.

As this initial fissure water moves so very slowly, the rates of solution are proportionately slow. But, steadily, the fissures do become wider, and as they do the flow rates increase and the solution and passage enlargement rates increase still further. Furthermore, not all the fissures enlarge at the same rate. Some joints may be larger than others and so capture a greater flow; openings above the shale beds commonly trap a large share of the downward moving water, and some openings are in the more soluble beds of limestone. These fissures then enlarge more rapidly than the rest, and, as they gain a hydraulic advantage in the flow network, they take more and more of the total flow and enlarge still more rapidly. These are the openings which eventually develop into sizeable caves. The remaining fissures of the network lose most of their flow to the larger, more efficient conduits, and they either dry out or just carry very slow moving water. They are relegated forever to the role of transmitting mere percolation water, and not the great cave rivers.

In the initial stages, all these fissures – microcaves in effect – are completely filled with water slowly percolating along them. But as they are enlarged they become more than capable of taking the normal flows of water feeding into them, and they can then develop in one of two different ways. Any fissure with a downhill exit will tend to partially drain out, so instead of being water-filled it has a free-flowing stream on its floor and an air-filled section above it. On the other hand, fissures which only drain out uphill, therefore under pressure, will remain full of water. In the broad view, the level of the drainage out of any fissure network is the spring, or resurgence, where the water flows out to daylight, normally near the floor of a valley at the edge of the limestone. The resurgence level controls the flooding level in the fissure system – below it all the openings are flooded, and above it they are freely draining. Hydrologists know these two zones as the 'phreatic' and 'vadose' zones respectively, and the boundary between them is the 'water-table'.

The water-table in cavernous limestone does not quite fit the conventional pattern of a water-table in normally porous rocks. In a sandstone or porous limestone it is a gently undulating sur-

Vadose canyons

The free flow of the stream in England's Easegill Caverns (above) has cut a meandering canyon in the cave floor. The same has happened in the Piaggia Bella cave streamway (below) in Italy, but there the stream has cut down through the limestone into a beautiful green slaty rock

face, inclined downwards, under the hills, towards surface rivers where it meets the ground. The river surface is part of the water-table, and the slope on the water-table provides the head of pressure to drive the groundwater through the rock. In cavernous limestone the water-table is almost horizontal, as very little head of pressure is needed to drive the water through the open cave conduits. Even more significant is that the cave streams in the limestone do not just flow on the water-table surface, as in the conventional model; this is due to the essentially watertight nature of the limestone, so that each cave stream is a system of its own. Consequently active, or even flooded, stream caves can pass directly over dry caves, which have been abandoned by other streams at lower levels. In South West Ireland, the River Aille on the surface passes right over a large dry cave passage (which cavers can walk along) – due to the fact that they are hydrologically independent units. Also a cave stream may pass straight down through the water-table, though the passages will be water-filled in the lower, phreatic zone.

The cave passages in the vadose zone may initially have been formed along a network of joints which include one or more uphill sections. The result is a U-tube in the passage, and this will remain flooded – a short phreatic section within the vadose zone, not necessarily related to phreatic sections in any other nearby cave passages, as each passage is unique. So caves relate to the water-table in that those below it are all full of water – phreatic – while those above it are mainly vadose but may include short phreatic sections.

Both phreatic and vadose cave passages initially develop from the same series of fractures – so for both types the overall patterns are dictated by geological structures. However, in detail, they are quite distinct from each other and may be recognized even when their streams have abandoned them.

Vadose caves are distinguished by their continuously downhill gradients and their lack of roof erosion. By definition they are formed by free-flowing streams with an air surface above the water. The streams erode the floor of the caves, cutting deep canyon-shaped passages, characteristically beneath a roof showing little sign of erosion – it is often just a smooth bedding plane of the limestone. The width of these vadose canyon caves is dependent upon the size of the stream, and the depth depends upon the rate of downcutting and the length of time it has been forming. Some of the canyon passages in the entrance series of the great Berger cave in France are 100 feet high and barely a yard wide. In the caves of the English Pennines some of these vadose caves are over fifteen feet wide but little over a foot high. Probably the largest of all the canyons is the Skocjan cave main passage in Jugoslavia – an immense 300 feet high and 150 feet wide.

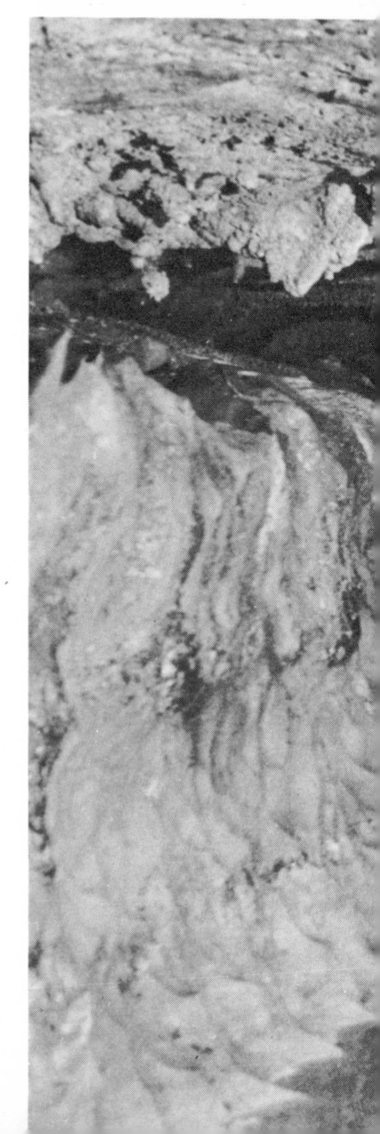

The continuous downhill gradient of vadose caves is again a feature of the free-flowing streams. The caves may have a sloping profile – as is common in inclined beds of limestone – or a stepped profile. Stepped profiles are most often found in beds of limestone that are lying nearly horizontal, where the caves have almost level stretches along the bedding planes, linked by nearly vertical shafts down the joints; this pattern is particularly well developed in the caves of the English Pennines. The shafts are formed by waterfalls, and in some parts of the world they range in depth to more than 1000 feet. These waterfalls set up great wind-currents in the shafts, which blow spray on to all the walls, and the main solutional enlargement of the shafts is by the films of spraywater flowing down all the walls. Consequently their most common form is nearly cylindrical, though a stream cut notch may lie down one wall. A series of shafts, connected by

The horizontal limestone beds in the English Pennines have been responsible for the staircase profile of many of the caves: the shallow canyon streamway beneath the bedding plane roof (below) is in Hesleden Cave, and the waterfall shaft (left) is in Lost Johns Cave

only very short lengths of level passage, give a very steep overall profile to a cave, and this is then commonly known as a pothole; the Epos Chasm in Greece, 1450 feet deep, but less than 100 feet long, is a fine example.

In complete contrast to vadose caves, the water-filled nature of the phreatic caves results in such diagnostic features as eroded ceilings and switchback gradients. These are easily recognizable, even when the water has drained out to leave the dry and explorable caves we see today. Erosion of the roof, floor and walls all at the same rate, by a complete fill of water, tends to give a circular cross-section to the phreatic caves, and the tunnels of Peak Cavern in England are nearly perfect examples. On the other hand some beds of limestone may be more soluble than their neighbours so that phreatic erosion results in the elliptical profile so well developed in the passages of Mammoth Cave, U.S.A. In addition, joints in the ceilings of the phreatic caves have commonly been eroded out to give rounded roof domes, completely closed at their tops, in contrast to the open-topped roof inlets of vadose caves, formed by trickles of percolation water.

The ability of phreatic water, under a head of pressure, to flow uphill may give phreatic caves a switchback, or even three-

The Welsh cave of Ogof Hesp Alyn is a phreatic system only drained 50 years ago by mining activity. Most of the passages are gently undulating tubes (above), with sediment accumulated at the low points (far left) and the mud coating on the ceilings showing where air pockets are trapped when the cave is flooded (above left)

dimensional maze pattern. Almost certainly the finest example of uphill phreatic flow is the resurgence of Vaucluse in southern France. There the water flows up a rounded tube, about fifty feet in diameter, from a depth of at least three hundred feet, to flow away on the surface. Not far north of Vaucluse lies the Favot Cave in the Vercours region near Grenoble. This opens out into a cliff face and is now abandoned and completely dry. Its entrance passage is a beautiful tube about twenty feet in diameter, dead straight for over three hundred feet, and inclined down into the mountain at about 40° from the horizontal. It is clear from the pattern of the cave that the water used to flow up the inclined tube, towards the valley – it is a classic phreatic cave, in essence an abandoned Vaucluse.

The first half of the twentieth century saw a phase in the science of geomorphology when general theories were applied to all the natural processes of the Earth's surface. The origin of caves did not escape a number of attempts at such generalizations. Some theorists proposed that caves were formed by free-flowing water, above the water-table, and others postulated cave formation in the flooded, phreatic zone. Based on the very reasonable premise that maximum limestone solution should take place where there is maximum contact and interaction between water and air, the water-table itself was commonly cited as the locus of cave development. A successively falling water-table, as the surface valleys were cut deeper into the limestone, could then account for multiple levels in cave systems. However a detailed, deductive, study of caves shows they are formed in all conditions. Phreatic, vadose and water-table caves can all be formed, even at the same time; the general theories were all over-simplifications.

There has always been considerable dispute over the existence of water-table caves, partly due to the fact that they are rather difficult to recognize. In the case of nearly horizontal limestones, it is almost impossible to distinguish between caves formed at the essentially horizontal water-table, and those formed in a single soluble bed of limestone. But some caves, in regions of inclined limestones, do have level passages cutting across all the geological structures. The Demanova Caves in Czechoslovakia are fine examples – they have a series of long passages, each gently inclined with the same gradient as the adjacent surface valley and cutting across the steep limestone bedding. They were clearly formed at ancient water-tables related to the old, higher-level valley floors. The water-table has also controlled the development of short horizontal caves around the edges of polje lakes in Jugoslavia. But on the whole, water-table caves are rare.

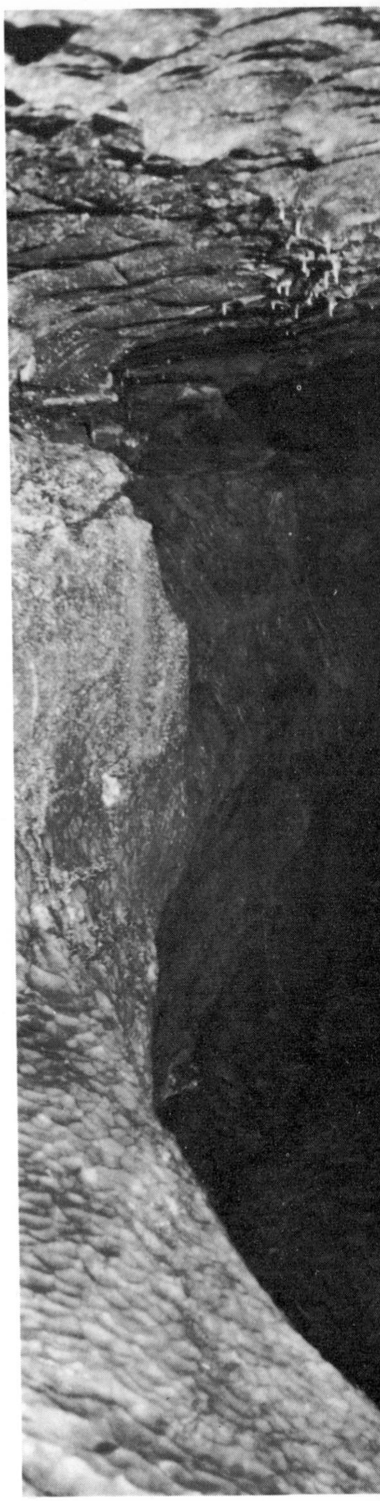

The Pennine caves of Kingsdale (right) and Lost Johns (below) both have vadose canyons incised beneath phreatic tubes

This is due largely to the dominant influence of geological structure in cave development; it is also due to the discontinuity of the water-table as a single surface – the hydrology is related to the conduit flows, in turn related again to the geology. What is more common in inclined limestone is a cave passage with a switchback profile, so that it oscillates up and down, by perhaps 100 feet, but keeps to an overall level zone. Swildon's Hole in the English Mendips is a classic of this type, where the cave developed down a bedding plane, then up a joint, down a bedding plane again, etc.; the cave maintained its overall level in order to utilize the maximum corrosion zone, just below the water-air contact (the water-table), but in detail it followed the geological weaknesses.

The overall result of erosion of the earth's surface is downcutting and surface lowering. Mountain peaks are eroded away, rivers cut their valleys deeper, and this lowering of river beds leads to a lowering of groundwater levels – phreatic zones are drained out so that they become vadose. In addition, vadose cave streams are continually obeying the laws of gravity in cutting down to lower levels. So, by these two processes, cave passages may be abandoned by the streams which cut them. Phreatic caves may be left completely dry or they may still carry vadose streams on their floors. And a long period of continued downcutting can leave a whole network of dry vadose passages, as the streams open up new caves via lower levels of joints and fissures. Consequently a long period of erosion in a limestone block results in a complex network of active and abandoned, phreatic and vadose, cave passages – the caves as we see them today. Indeed the only phreatic passages which anyone except a cave-diver ever sees are the abandoned ones. Over time, surface downcutting and consequent lowering of the associated water-table means that nearly all phreatic caves are eventually drained out and transferred to the vadose environment. So most visible phreatic caves are the old ones, and they are commonly intersected by younger vadose passages. The reverse – the flooding of vadose caves – is rare.

The general pattern of cave systems all over the world is therefore remarkably similar – old phreatic passages, intersected by the younger vadose passages; some of the latter are normally active and may feed into a modern, inaccessible level of active phreatic caves. The Trou du Glaz cave system in France is almost typical. Its overall depth is just about 2000 feet, while its total passage length is around twenty miles. About a third of the passages are abandoned phreatic, and just a few of these have younger vadose streams in them. Another third are abandoned vadose passages, old but not as old as the phreatic caves, and youngest of all are the remaining third which are the active vadose caves.

The effects of time and structure have led to a complex but fascinating network of caves.

It is clear enough that caves vary in age – but it is not so easy to calculate just how old they are. One feature that cave age does not relate to is the age of the limestone in which they are formed. Most of the cavernous limestone in Britain was formed in the shallow seas of the Carboniferous period about 350 million years ago, while most of the Alpine limestones are of Jurassic age, about 170 million years old. But in each case practically all the caves have been developed within the last two million years. Until that time the limestones were buried under great thicknesses of other, younger rocks, and cave formation only begins when the limestones are exposed on the surface – and subjected to contact with rainwater. Most of the European karst landscapes only date from the Pleistocene period – the last two million years – and consequently the caves are of the same age. The Pleistocene period included all the Ice Ages, and the glaciers of these colder periods were largely responsible for the cutting of the major valleys through the karst and the resultant change from phreatic to vadose conditions in so much of the limestone. The great contrasts in the caves we see today are mainly due to these unprecedented climatic changes in the Pleistocene era when the caves were forming.

Not all cavernous limestones are as old as those of Britain and the Alps. A truly remarkable region is the Pokhara Valley in Nepal. This has a floor of Pleistocene limestone – deposited only a few million years ago by rivers draining off the Himalayan peaks (themselves formed of older limestone). Changes in the local environment, noticeably the climate, have meant that the rivers no longer deposit limestone but erode it. Perhaps in only the last ten thousand years or so, great cave passages up to thirty feet high and wide have been cut by these rivers through the Pokhara limestones – the Harpan River Cave is just one.

The actual rate of cave formation depends on the chemistry of both the limestone and the erosive water. The latter is the more important point, and it in turn depends largely on both the climate and the vegetation, with its influence on carbon dioxide concentrations. Caves tend to form faster in tropical regions and the Harpan River Cave is an obvious example of a very rapidly formed cave. A more precise estimate of cave erosion rates has been made in the limestones of County Clare in Ireland. There, vadose stream caves about fifteen feet high have been shown (by their relationship to the general landscapes) to have formed since the last retreat of the Ice Age glaciers – about 12,000 years ago. This rate, of about one sixtieth of an inch per year, matches well with very accurate measurements made of the erosion of cave floors over periods of a year or so.

It is a peculiarity of erosive processes that, through them, nature tends to undo its own handiwork. The natural progression beyond the formation of caves is that they shall be removed, destroyed, or filled in. Caves frequently form underneath surface valleys, where there tends to be a concentration of groundwater flow. Subsequent deepening of the valleys, perhaps by glaciers, then completely removes these caves and their surrounding limestone. In some cases relic segments of passages are left in the sides of the deepened valleys. Short remnant caves left in this way and now perched high in the ridges of karst areas in the Alps and Rockies just indicate the extent of the great cave systems removed during the Ice Ages.

Collapse is another mechanism of cave destruction. If the surface is lowered by erosion until caves are only just below ground level, the layer of rock above them may be so thin that their roofs may collapse. Even more important is the collapse of cave roofs far below the surface. This is not a rapid process – few people have seen even a single rock fall from the roof of a cave; but percolating water eating out the joints in the limestone does, slowly, loosen blocks in the ceilings of cave passages. In very old cave passages even this slow rate of collapse can result in massive rubble heaps on their floors, particularly where there is no longer a stream to remove the fallen debris. Collapse is frequently cited as the cause of large cave chambers. But this is incorrect – in fact the real cause is quite the contrary. A mass of solid limestone from a cave roof will occupy a far greater volume when it is a loosely packed heap of rubble on the cave floor – collapse does not form caves, it fills them in. Some chambers do have obviously broken roofs and collapse blocks covering their floors – but they can only have been formed by collapse into even larger, previously formed chambers.

Caves are also filled in by material carried in from outside. A change in climate, or the partial loss of water to another newly formed passage, can significantly reduce the flow in a cave stream. It may then be that the reduced stream is no longer able to erode and transport sediment but instead deposits it in the caves. Hence the huge deposits of sand and mud in so many caves. Not only is such sediment a characteristic of an abandoned cave, but it is also indicative of changing stream flow conditions on the surface. The Ice Ages of the last few million years have caused immense variations in stream flows and there have been many periods of sediment deposition. Yet again the glaciations can be seen to have influenced cave detail, not only in the patterns of their formation, but here in the means of their destruction and blockage as well. Similarly the glaciations influenced the transport and deposition of material carried by the streams in solution – the calcium carbonate itself. And the

Underground architecture

The main passage of Dow Cave (far right) in the English Pennines has a wide nearly flat bedding-plane roof; but in the nearby Lancaster Hole, Montague West Passage (below right) has an arched roof as it is a phreatic tube partly filled with sediment and stalagmite. Another Pennine cave, Out Sleets Beck Pot (right), has a scalloped rock bridge left across the passage as this stream has cut down through the jointed limestone. A chaotic chamber in Shatter Cave, in the English Mendips (below), was eaten out of the rock when full of water, and has since been decorated with calcite deposits

Yordas Cave (below) in the English Pennines has a level floor of stones, sand and clay in its main chamber. But the inlet passage in France's Pierre St Martin Cave (opposite) has great banks of mud and boulders sloping down towards the central trench

deposition in the caves of this mineral gives the huge variety of stalactites and stalagmites – in this context merely another means by which caves are filled in.

The form of a cave as it exists today is therefore the result of a whole range of different, sometimes opposing, natural processes. The Gaping Gill cave system, six miles long and one of the finest in the English Pennines, is a very old cave, and a fine example of this multiple origin. It contains a maze of ancient, abandoned phreatic tunnels. In addition there is a whole series of younger active vadose stream caves, and a modern inaccessible flooded zone. There is a massive chamber heavily modified by collapse and with a sediment-covered floor – it is impossible to guess its original dimensions. The old phreatic tunnels are partly filled with great volumes of gravel, boulders, sand, clay, stalactites and stalagmites. Exploration along the various arms of the system is halted in some caves by breakdown heaps, in others by sand-fills reaching the roof and in others by stalagmites blocking the way. Yet others end in heaps of debris only a few yards from the surface – the continuations of these passages were removed long ago when glaciers scoured out the valleys. It could be said that half of the Gaping Gill caves are destroyed or filled in – the system is smaller than it once was. Yet the modifications have added to its complexity and variety, and the result of millions of years of both erosion and deposition is a typical and fascinating system of caves.

6 Decorated Caves and Cave Deposits

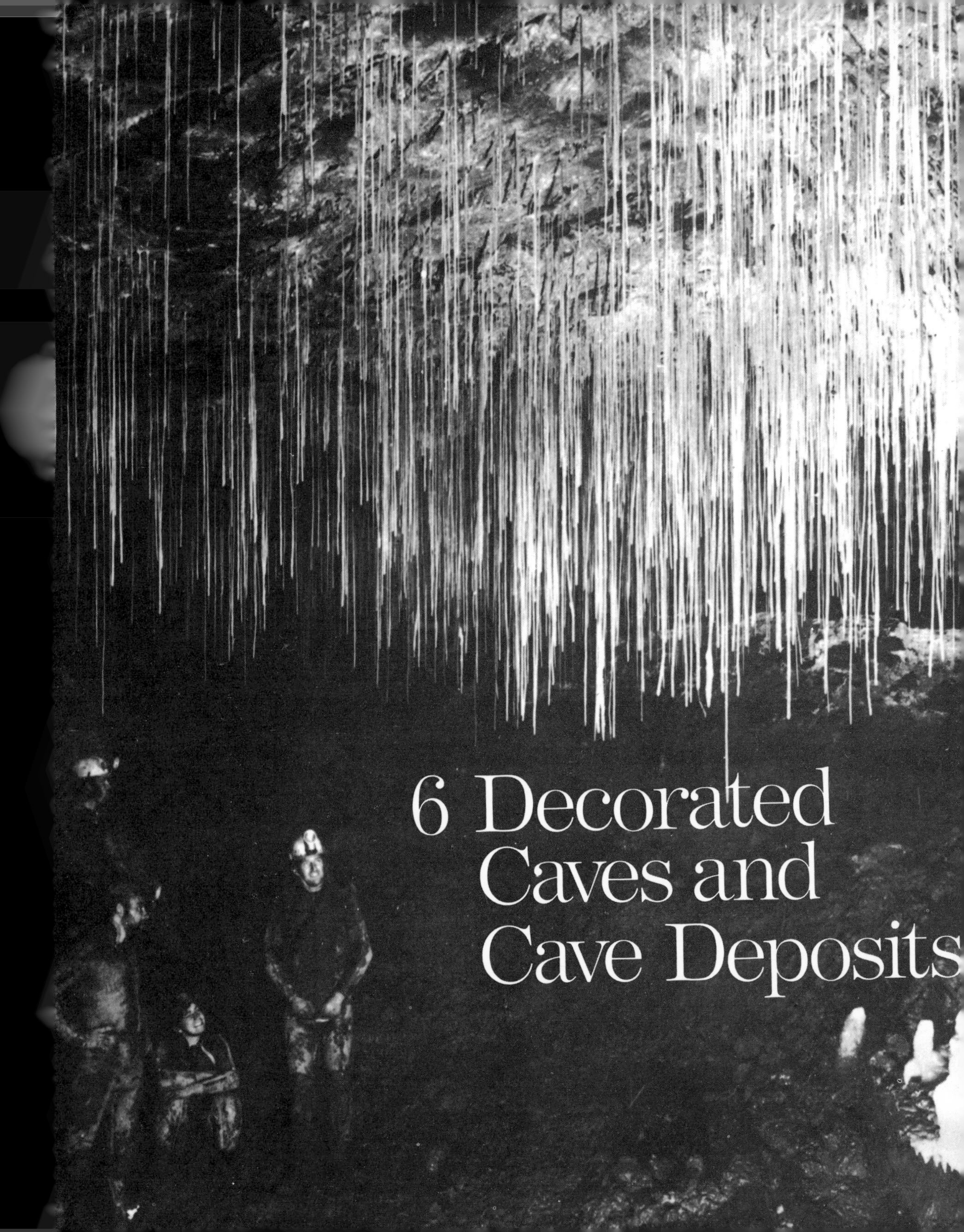

Balch Cave had only been discovered for about ten years when it was almost completely quarried away. But in that short time it became widely known as a fabulous decorated cave. Situated in the English Mendip Hills it was formed in limestone, but any visitor to the cave saw very little limestone once inside. For the walls, ceilings and floors were nearly all covered in sparkling white calcite decorations. Stalactites hung from the ceilings and festooned the walls, flowstone covered the passage floors and the chambers were a wonderland of stalagmites and crystal pools. Only the musical sound of drips of water broke the silence in this beautiful cave; the rivers which once formed it were long gone. The little water it had was not enlarging the passages – it was filling them in with the whole spectacular range of calcite deposits. So what was special about Balch Cave? What is the difference between the world's decorated caves and those clean-washed caves containing active streams and rivers?

Basically, the answer is time. A valley in a young mountain range is carved out and deepened by the stream which flows down it. But a valley in an ancient, eroded down, lowland area tends to be filled by sediment deposited by its river – to form a floodplain, for example. Time results in a change from erosion to deposition. The same happens in the caves.

Percolating water, and then a stream, form a cave. But even in so doing the water changes its own environment, and over the passage of time it makes further changes. Eventually, because of these changes, the water ceases to erode and enlarge the caves, and starts instead to deposit and fill them in. It is a natural process that once a cave has been formed it will be at least partly filled in. This is part of nature's continued efforts to even everything out – at first the limestone is dissolved away, because it forms a hill area; and then the caves must be filled in, because they are spaces in solid rock. Fortunately, the caves do not just fill in again with limestone, except by collapse. Instead they gain a whole variety of deposits ranging from sand and mud to a number of specific minerals. Of the latter, calcite and aragonite are by far the most important – they form nearly all stalactites and stalagmites.

Two main types of water enter cave systems. First there is the water that collects into streams on the surface and enters the major sinkholes – this does most of the cave erosion, due to its large volume. But very much more corrosive in its attack on the limestone is the second type – the percolation water. This just seeps into narrow fissures in the rock, and in most parts of the world it does so having already filtered slowly through a soil cover. And there it gains its corrosive power, in the form of large concentrations of carbon dioxide. Water will always contain an amount of carbon dioxide which is in equilibrium with that

Deep inside Easegill Caverns in the English Pennines there is a magnificent decorated chamber, Easter Grotto (above). A forest of straws hang over pure white stalagmites, and closer views (right and centre) reveal how some of the straws curve away from the vertical and how others are covered in little branching helictites. The straws in nearby Pippikin Hole (far right) hang above a glittering floor of flowstone

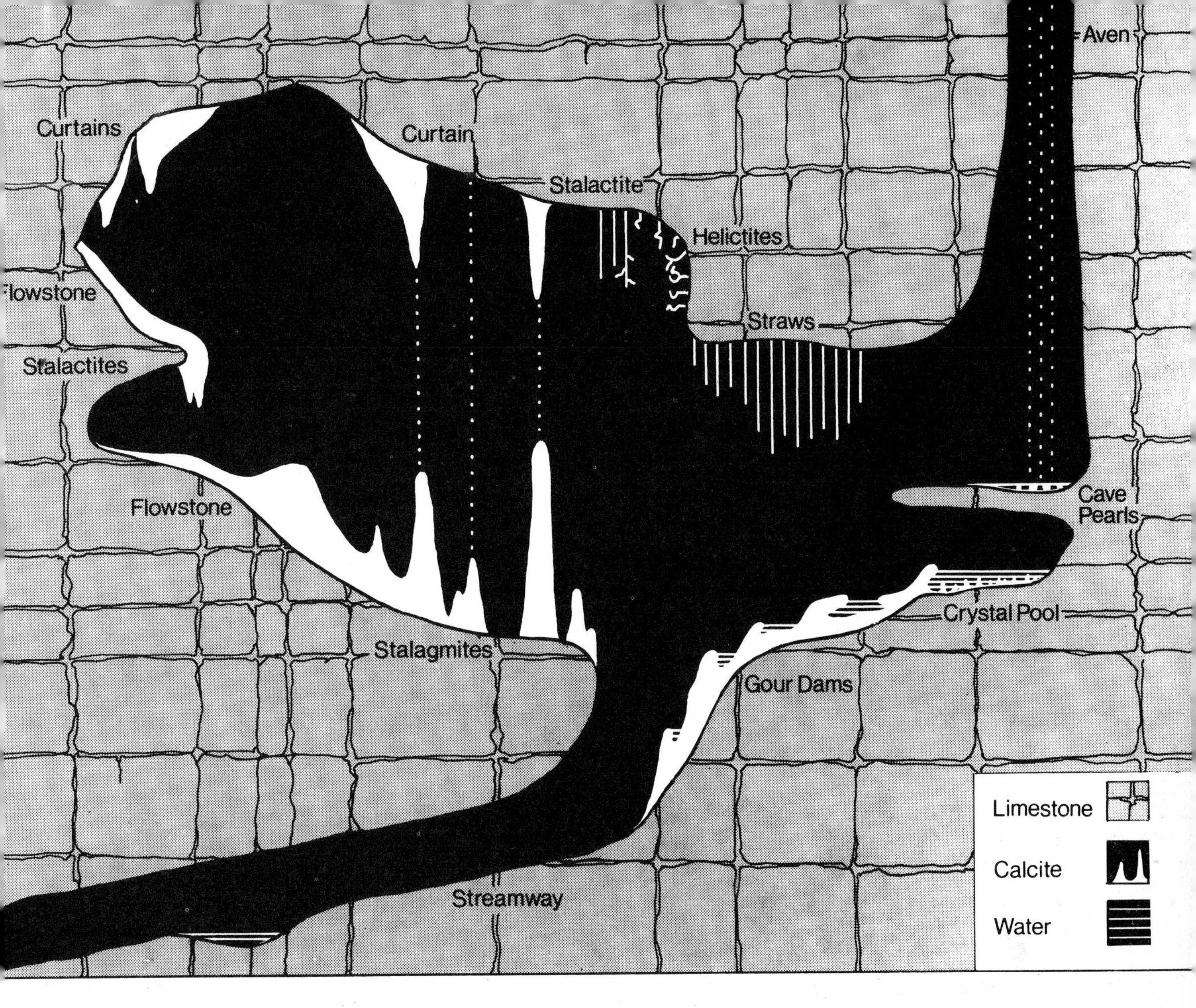

Cave calcite is deposited by water in a variety of different forms

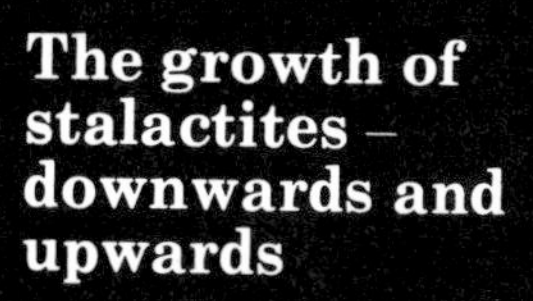

The growth of stalactites – downwards and upwards

France's Coufin Cave (below) contains a hanging forest of straw stalactites, each a hollow calcite tube the diameter of a drop of water. It is in the helictites that water must somehow flow uphill: in Withyhill Cave in the English Mendips they are the same diameter as the straws (far right), but in Lebanon's Jeita Cave they are dwarfed by the drop of water that hangs from one (near right)

Cascades of stalactites and flowstone hang down the wall of the Gournier Cave (far left) in southern France. Nearby, the Berger Cave has a dramatic fringe of stalactites in one of its chambers (above left) while a mass of calcite curtains hangs from the roof of the Père Nöel Cave in the Belgian Ardennes (below left)

The columns (above) in Wales's Ogof Ffynnon Ddu cave are a beautiful group of tall stalagmites, some of which have grown to make columns meeting the cave ceiling. Shatter-hill Cave in the English Mendips contains an unusually pure calcite curtain (right) which is dazzlingly white

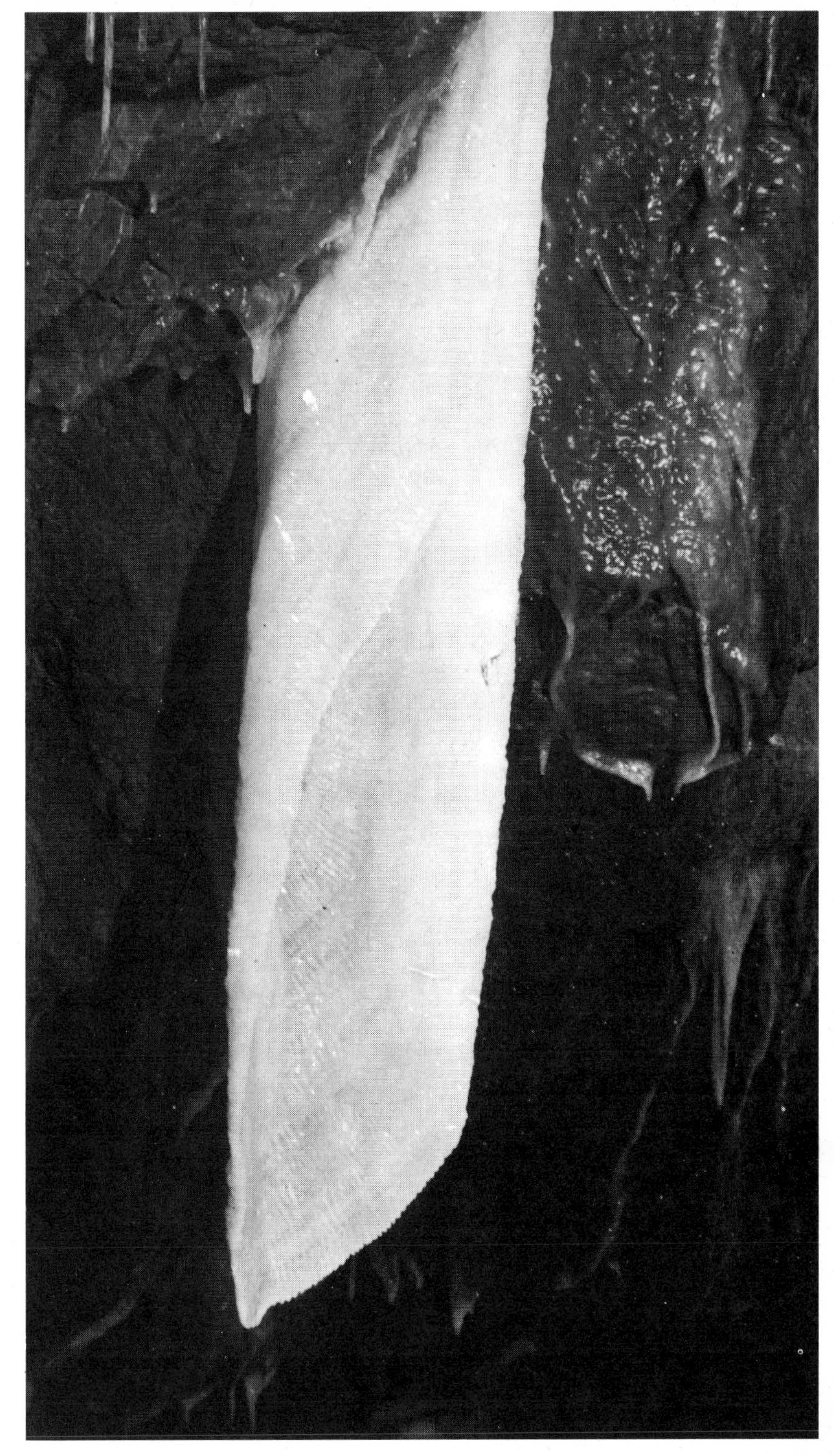

in the air in contact with the water. Soil atmospheres have very high contents of carbon dioxide, due to the organic activity, and so soil waters have similarly high concentrations. The water therefore enters the limestone capable of dissolving large quantities of rock.

As the water percolates down through the limestone it steadily takes the rock into solution – perhaps reaching concentrations of calcium carbonate in excess of 300 ppm, when the carbon dioxide and calcium carbonate are themselves in equilibrium in the solution. Further down, this slow flow of percolation water may drip out of its joint into a ready-made cave. But there it meets the cave atmosphere, which is very similar to normal 'outside' air and contains only low concentrations of carbon dioxide. So carbon dioxide diffuses out of the water into the air – to regain equilibrium. But the amount of calcium carbonate dissolved in the water depends on the carbon dioxide concentration, so it too must diffuse out of the water to maintain the gas-mineral equilibrium. The calcium carbonate precipitates out on the wall of the cave – forming stalactites and stalagmites, or speleothems as they are collectively known.

The volume of these deposits depends on the carbon dioxide loss which the water suffers. As cave airs have a relatively uniform composition this therefore depends on the amount of gas originally dissolved in the water – in turn related to the carbon dioxide content of the soil atmosphere which it passed through. This depends on the level of organic activity in the soil, and this is controlled by the climate. Consequently caves lying beneath thickly vegetated tropical surfaces tend to have more and larger speleothems than caves existing beneath barren arctic landscapes – in fact caves in the colder regions are commonly almost devoid of stalactites.

The two minerals precipitated in caves by these percolating waters, calcite and aragonite, both consist of calcium carbonate. They are polymorphs – minerals with similar compositions but contrasting crystal structures. Theoretically aragonite should not be stable in the pressure and temperature conditions of caves but it appears that the presence of minute quantities of other elements allows aragonite to be deposited instead of calcite in some cases. Calcite is however still the most abundant mineral precipitated by cave water. Furthermore it is very difficult to distinguish between aragonite and calcite except in a laboratory, and many examples of aragonite deposits probably go unrecognized – the term 'cave calcite' is frequently used to describe deposits which may in fact consist of the two minerals.

The simplest of all the many forms of cave calcite is the straw – a type of very thin hollow stalactite hanging down from a cave roof (the stalagmites grow up from the floor). A single drop of percolation water seeps out of a microscopic joint in the limestone and hangs on the cave roof for a time before it falls to the floor. Carbon dioxide diffuses into the cave air, and a ring of calcite is deposited around the edge of the water drop where it clings to the rock. The drop falls away and another forms in its place so that the process is repeated, with each drop of water hanging onto the end of the calcite precipitated by the previous one. The result is a long cylindrical straw with the water feeding down its inside. It has an almost constant diameter because of the constant size of a drop of water. Its diameter and wall thickness are very similar to those of a drinking straw – hence the name. If the conditions are right, straws tend to form in large numbers. The show-cave of Choranche (the re-named Coufin Cave) in the French Vercours has tens of thousands of straws hanging from the ceiling of its main chamber. Each is about six to ten feet long. It is only rarely that straws grow longer than

The drop of water hanging on the end of a cave straw has tiny calcite crystals (below left) growing inside it. Eventually the stalactite will be linked to the stalagmite (below) by continued deposition from the drop of water. Both these formations are in Easegill Caverns in the English Pennines

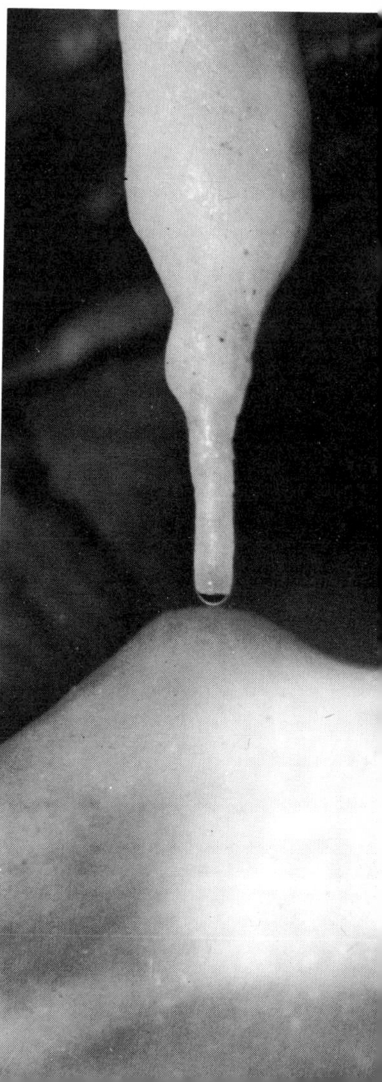

this before they break off under their own weight. One of the longest known straws is in a small chamber in Langstroth Pot in the English Pennines – it is very nearly fifteen feet long; the world record probably belongs to a straw 20·4 feet long in Boranup Cave in western Australia.

The slender uniform shape of the straws is due to water only flowing down their insides and meeting the cave air for the first time at their tips. Should water flow down the outside of a straw, it will continuously diffuse carbon dioxide into the air, and will deposit a larger layer of calcite – to form a stalactite. Some stalactites do have a central canal, like in a straw, but others do not and must therefore owe their entire growth to water flowing down the outside of the structure. Stalactite shapes vary immensely, due to the varying rates of water supply and diffusion of the carbon dioxide into the air. The simplest, and commonest form, is a slender cone tapering down to a point. This develops

In the Père Nöel cave in Belgium (right) the stalactites and stalagmites are covered with a mass of tiny helictites, but in England's Darnbrook Pot (far right) the calcite pillars stand clean and alone in the rocky passage

because the maximum deposition of calcite is where the water first meets the cave air – at the top of the stalactite. As the water then flows down the stalactite surface it steadily approaches chemical equilibrium with the air, and so deposits progressively less calcite. A section cut through a stalactite shows that it consists of concentric layers of calcite, successively deposited by the moving film of surface water. The more complex stalactite forms mostly owe their origins to changes in precipitation rates – and these normally relate back to climatic changes; the range of varieties is almost endless. The low tensile strength of calcite places a limit on the size of stalactites. Although they are probably the commonest cave deposits, found all over the world, really large examples are rare. There is just one stalactite hanging from the roof of the dark and sombre chamber in the Pollanionain Cave in Ireland; but it is thirty-eight feet long, probably the longest in the world.

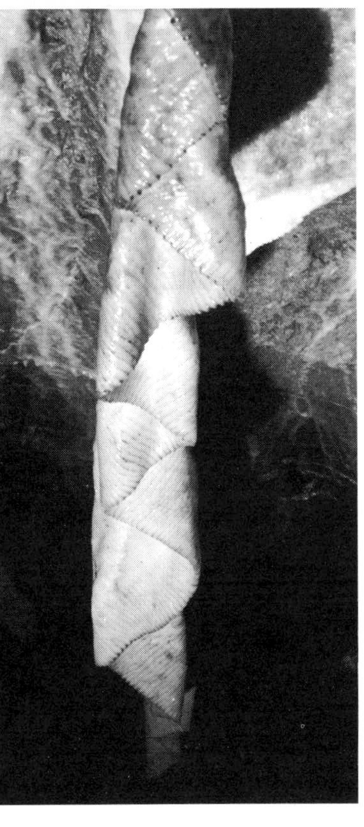

A cluster of sparkling white stalactites are set in a corner of Withyhill Cave in the English Mendips (above left). Much larger are the stalactites down the wall of the main chamber in G.B. Cave, also in the Mendips (left). In Père Nöel Cave, Belgium, the dripping water has formed a great veil of calcite curtains (right), and a single curtain in Withyhill Cave (above) shows how it is banded and folded

Not all the water on cave ceilings just drops from discreet points. It may run down the ceiling slope – depositing a thin streak of calcite as it does so. This deposit then forms a low edge on the ceiling, and following drops of water will flow down adding further to the deposit. The result is a curtain – a great sheet of calcite hanging from the cave roof; many examples are several feet high and wide, whereas they are commonly less than an inch thick. Successive layers, each roughly parallel to the cave roof, may have contrasting colours; white or yellow-brown due to varying iron content of the solutions are the commonest, and the results are often graphically known as 'streaky bacon'. More commonly curves and folds in the calcite sheet heighten the analogy to a flexible curtain. Curtains are formed in caves all over the world – wherever there are sloping ceilings. In Britain the finest examples are in St. Cuthberts and Withyhill Caves in the Mendip Hills. Curtains are common in so many of the Mendip caves because there the limestone beds are inclined at about 45° – giving plenty of suitable surfaces for their formation. In contrast, the Pennine caves are in nearly horizontal limestones and curtains can only be formed on erosional undercuts in some of the cave passages.

Weirdest of all cave deposits are the helictites. Fragile, cylindrical, calcite forms, they curve and twist in every direction – up, down and sideways. Some thicker than a man's finger, others less than a tenth of an inch in diameter, their eccentric shapes completely defy gravity. They grow out from vertical walls, sometimes in lines along hairline cracks in the limestone, but also up and down from level floors and ceilings; others grow from the sides of stalactites and straws and some have straws growing from them. Helictites are surprisingly abundant in caves all round the world, but most are so small they are easily missed – not many exceed four inches in length. Timpanagos Cave in the Wasatch Mountains of the U.S.A. contains thousands of helictites in great tangled masses – probably one of the finest collections in the world. But no one knows how helictites are formed. They don't fit in with any of the easily explainable means of mineral deposition, and many special theories have been proposed to account for their growth. The more acceptable theories have mainly depended on organic influence, generally by algae, interruptions in the calcite crystal growth, wind currents to distort normal stalactite growth, or capillary action to give an upward supply of water. In some caves whole groups of helictites point in the same direction – and these may well be due to deposition from drops of water blown around to the leeward sides of stalactites. But the vast majority of situations do not fit such a simple pattern. The most acceptable theory is a composite one – where the eccentric shapes are defined by irregularities in the atomic structure of the calcite, resulting in off-set crystal growth, with the supply of water passing uphill where necessary due to capillary action.

All this dripping water, from stalactites, curtains and helictites, eventually falls to the floors of the caves – to form yet another series of calcite deposits. These are mainly an infinite variety of stalagmites, as well as various types of flowstone. They are developed because the dripwater arrives on the floor still over-saturated with calcite with respect to the cave air. Not all water has time to reach the equilibrium state while it is hanging in drops on the cave roof. More carbon dioxide is lost as the water falls and water may also evaporate at the same time – further concentrating the solution – so that calcite is precipitated as soon as the drop of water hits the floor. If the floor is level, the drop of water on contact spreads out and flows away in all directions, precipitating calcite until it is in equilibrium with the air. So a disc of calcite is formed. But in detail this will be a very flat cone, for the rate of deposition falls off both in time, as equilibrium is approached, and also outwards as a greater area is covered. So dripping water over thousands of years can build up successive conical layers to form stalagmites of all shapes and sizes.

Some of the most beautiful stalagmites are the slender towers with almost constant diameters. There is a magnificent group of them in the Columns Hall in Wales' Ogof Ffynnon Ddu cave; some have reached the passage roof, to form columns about nine feet tall, while others are still free-standing stalagmites. They are almost perfect cylinders, yet are nonetheless made of cones of calcite sitting on top of each other. In this case, the water reached equilibrium with the air after it has flowed down only a short part of the stalagmite; it then flows down the lower part without depositing any more calcite – so there is a limiting diameter to the stalagmite. In some places there is much more calcite to be deposited from the dripwater, and it spreads further. The result may be great sheets of flowstone – massive layered

Helictites grow from fragile cave straws in Easegill Caverns in the English Pennines (below left). In Withyhill Cave in the Mendips much finer and more delicate helictites grow from a cave wall (above top) and from a small stalagmite pillar (above). The helictites of Timpanagos Cave near Salt Lake City grow in profusion on a mass of stalactites and curtains (above left)

calcite all over a cave floor. Intermediate between the flowstone sheet and the stalagmite is the massive conical hump of calcite known as a boss. These can grow to immense sizes; there is one in the Favot Cave in the French Vercours which is over fifty feet high and wide.

But again, few stalagmites are uniform columns or cones. Those in the Carlsbad Caverns in the U.S.A. are among the largest and best known in the world. They vary very much in shape, but nearly all are clearly composite forms – individuals vary in diameter over their height and their external structure shows layering. The composite forms are due to change in deposition rates. The cylindrical form of Ogof Ffynnon Ddu's stalagmites depends on a uniform deposition rate; but a climatic change, followed by more calcite in solution, can give a stalagmite that thickens upwards. Commonly the lower parts are then thickened by the addition of curtains and other stalactite forms. More changes and more layers, and the stalagmites become as complex and massive as those in Carlsbad. A decrease in deposition rate was responsible for the massive stalagmites with little columns on top which are so common in Jugoslavia's Postojna Cave. Just under a hundred feet is probably the record height for a stalagmite – and this goes to the tallest of a whole group in a fabulous chamber in France's Aven Armand. The stalagmites of Armand are a rather unusual variety – they appear to be made of rounded, irregular, hollow cones, which are concave upwards. This is probably related to the deposition from drops of water breaking up as they fall from great heights, or at the edges of small pools on the tops of the stalagmites.

The palm-tree form of the Armand-type stalagmites, and the mysterious shapes of helictites, would have been good support for the theories popular in the first half of the eighteenth century, which suggested that stalagmites were alive and grew like plants. (Before then, cave formations had been thought to be 'petrified' water, an idea first proposed by Pliny in the year 77.) It was only in 1771 that the solution of calcite in water was recognized – and the growth was seen to be due to deposition of layers. An average rate of growth for calcite stalactites is now considered to be somewhere around an inch in every century. The rate obviously varies over different parts of the world – it is fastest in the tropics due to the high solution rates beneath the organic soils. On the other hand, the relative abundance of stalactites and stalagmites depends, not on solution rates, but on the flow-rates of the dripping water. A low rate gives a drop of water a long time hanging on the ceiling, and so gives more stalactites. The converse results in the water falling off the ceiling before much deposition has taken place, and more growth of floor deposits. However that cannot be the whole story; there are cave

Stalagmites range in shape from amazing tiered forms such as in Carlsbad Caverns USA (left) to the slender towers like that in Père Nöel Cave, Belgium (above)

The flowstone floor in the Naciemento del Rio Cave in Spain has been formed by a mass of tiny gour pools (above left). In Withyhill Cave in England (above) the calcite flow is re-crystallized and now sparkles in any light. Calcite straws hang beneath a remnant of false floor in Withyhill Cave (left); a short-lived pool must once have submerged part of the straws as indicated by the clustered calcite growth on the tips

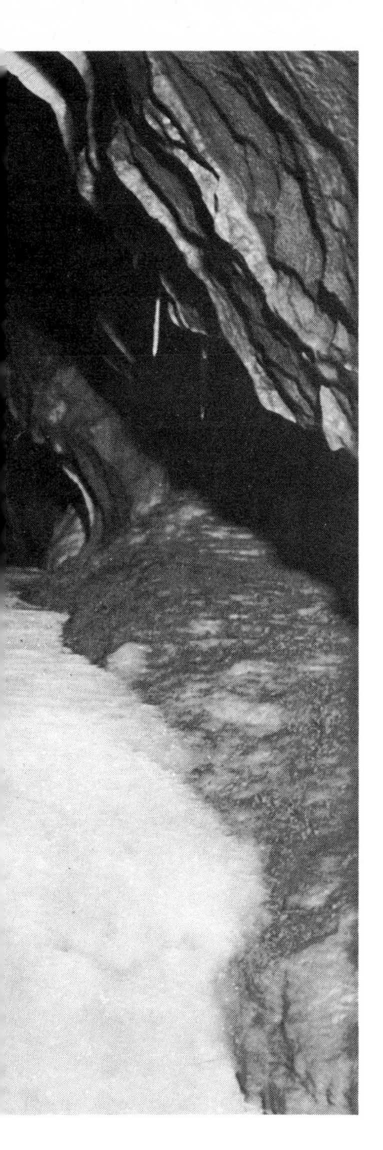

passages in the English Pennines which have sections containing just stalactites right next to sections almost exclusively decorated with stalagmites.

Floor deposits do not end with stalagmites: pools of standing water have their own types of calcite deposition – dripwater landing in pools can never form stalagmites. Instead the water in the pools becomes over-saturated, as further carbon dioxide is lost, and the over-saturation is further aided by evaporation of the standing water. Crystals of calcite may therefore grow in cave pools and they can be very beautiful. The crystals may completely line the floor and sides of the pool. Such pools, now dried out, are to be found in Shatter Cave in the English Mendips; they are ringed with platy crystals nearly four inches across. Other cave pools in England contain sharply pointed pyramid crystals, though these rarely exceed an inch or so in length. This crystal growth tends to be fastest at the surface of the pools, where there is the maximum diffusion and evaporation loss, so that calcite sheets may develop over the water. Czechoslovakia's Demanova Caves have some fine examples of these crystal surfaces.

A continued supply of water to a pool results in overflow at the lowest point on its rim. As this thin sheet of overflow water crosses the pool rim there is even more scope for evaporation and diffusion, so that more calcite is deposited. And so the pool builds its own dam – known as a 'gour' barrier. These can be immense – thirty feet or more high and containing pools which may be three hundred feet or more across. A staircase of gour pools provides one of the finest spectacles in caves. In France's Berger cave, the Hall of Thirteen has long been famed for its fabulous calcite decorations. Its entire floor, fifty to sixty feet wide, is crossed by curved gour dams which form a natural staircase down more than 200 yards of the passage. The slopes on the dams vary, some are nearly vertical, some are very gently inclined – almost like flowstone banks. Some of the smaller gour pools are lined with crystals, but much more common are just masses of rounded concretionary growths of calcite.

These concretionary calcite growths also occur all over the walls of some caves, and they indicate that the cave has been temporarily flooded by calcite-saturated water. Rather misleadingly, the name 'cave coral' is normally applied to this type of deposit, though the American name, 'cave pop-corn', is rather more descriptive. The U.S.A's Carlsbad Caverns contain some of the most extensive cave coral deposits – many of the vast stalagmites and stalactites in these huge chambers were temporarily submerged, so that their surfaces are now covered with these concretionary lumps. One beautifully tapering stalactite, about six feet long, must just have had about eight inches of its

Needle-sharp calcite crystals (circled above) grew in a lime-saturated pool that once existed on the floor of Easegill Caverns in the English Pennines (above). Similarly saturated water disturbed by falling drips forms no crystals but instead a mass of spherical cave pearls; those in Castleguard Cave, Canada (far right) are up to an inch in diameter

tip dipped in this bygone pool of standing water. Concretions therefore developed, almost in a cluster, on this submerged section, and the whole formation is now most aptly named the 'Lion's Tail'.

A very special type of concretionary calcite forms in a pool of saturated water continually disturbed by dripwater falling into it. Then calcite tends to coat minute pebbles or sandgrains on the floor of the pool, and the agitation of the water results in movement of these grains so that the calcite is deposited right round them in a complete shell. Continued deposition of this type results in spherical forms, known as cave pearls, built up of many layers of calcite. Pure white cave pearls, sometimes an inch in diameter, can occur, usually in 'nests' where a number are formed in one pool. To form pearls the pool has to be only an inch or two deep, and must lie beneath a high roof aven or dome which has a steady drip of water falling free down its centre. Very fine nests of pearls have been formed in St. Cuthberts and the Gaping Gill caves in England, and in Canada's Castleguard Cave, and most lie on the floors of shafts at least eighty feet high.

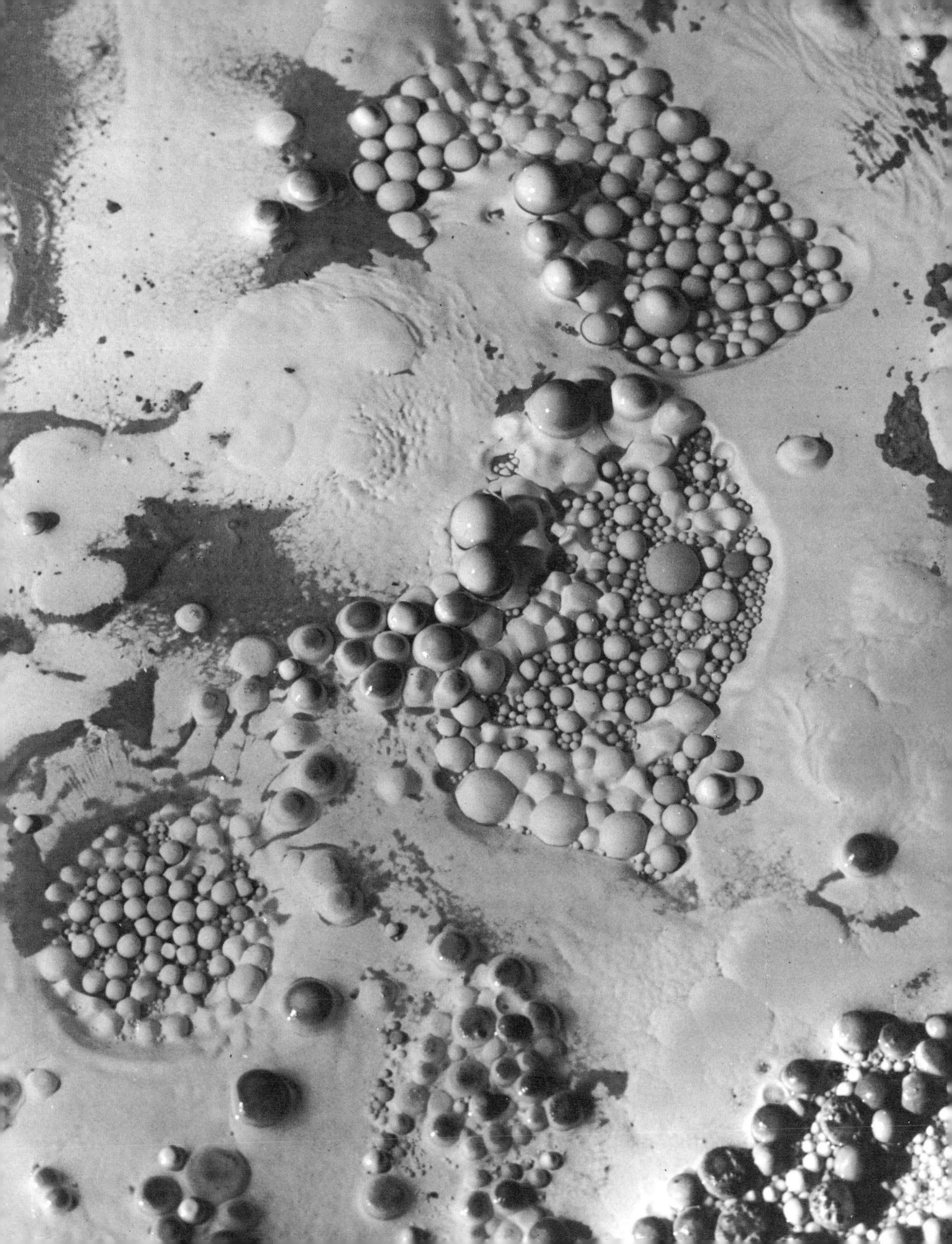

Unusual mud forms cling to the wall of St Cuthberts Cave in the English Mendips

Calcite forms the greatest variety of cave deposits, but does not have a monopoly on the decorations found in caves. In total quantity, sand and mud deposits are more important, but they rarely develop the spectacular shapes of cave calcite. An exception is the mud stalagmite, formed by mud-laden water dripping off a ledge. Even they are not common, and are usually rather short and fat – rarely more than eight inches tall. But there are strong competitors to calcite in providing the most beautiful cave decorations – the sulphate minerals and ice.

A whole variety of sulphate minerals has been recorded in caves, but by far the most common, even in limestone caves, is gypsum. It consists of hydrated calcium sulphate, is highly soluble in water, and is easily precipitated when percolation water, saturated with it, evaporates as it seeps into an open cave. Lying straight above the limestones which contain America's immense Mammoth Cave system is a sandstone containing quite significant quantities of pyrite – iron sulphide. Oxidation of this sulphide, in the groundwater, gives sulphates which then react with the calcium in the nearby limestone to form gypsum. Conse-

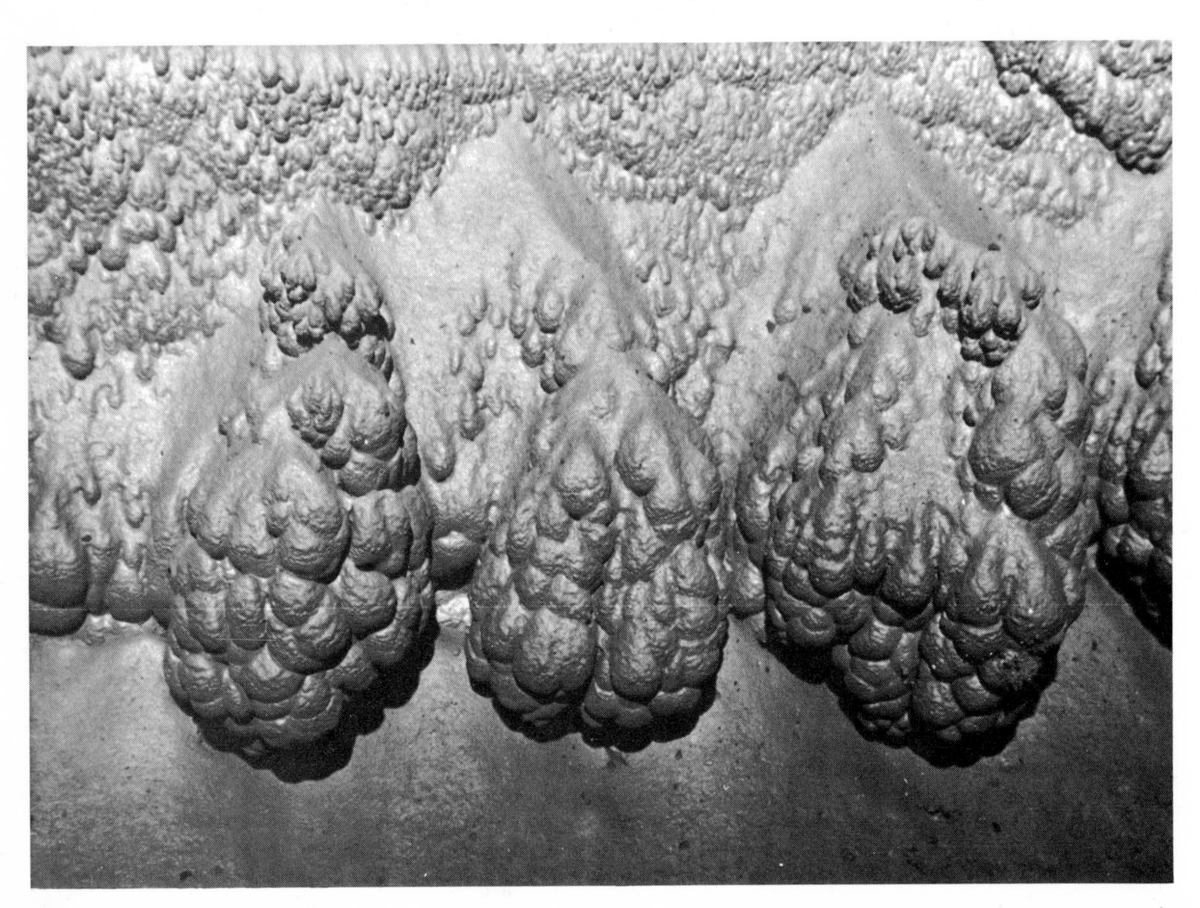

Stalagmite forests

The Berger Cave, in France, is famous for its stalagmites. The ranks of stalagmites bordering a calm lake (below) are tall smooth pillars, but the smaller individuals nearby (right) are complexly tiered

Lakes that build their own dams

Water trickling down this passage in the Little Neath River Cave in Wales (below) continuously deposits more calcite on each of the tiny gour dams. A single deep gour pool in France's Gournier Cave (far left) is surrounded by calcite flowstone, and the nearby Berger Cave (left) contains a veritable staircase of gour dams and pools

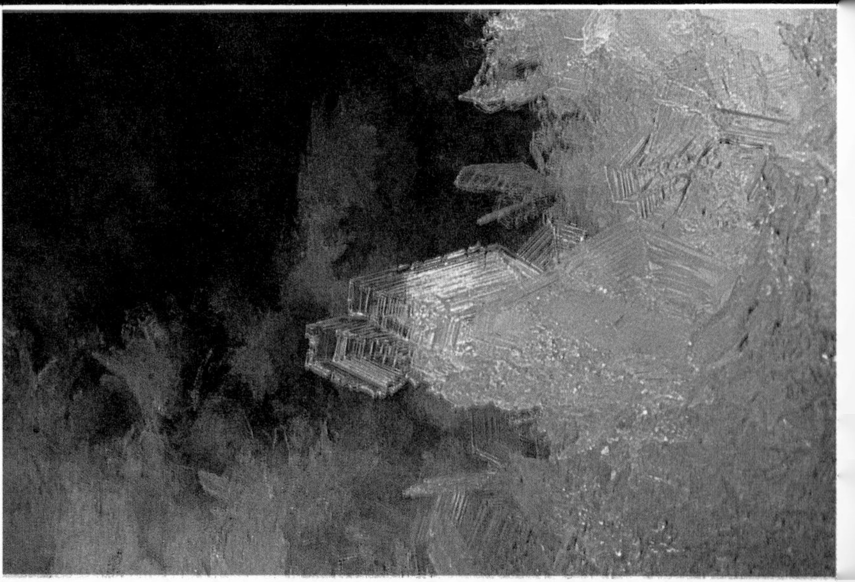

High in the Canadian Rockies, the walls and floor of Plateau Mountain Ice Cave are completely covered by thick layers of ice crystals – most of them hand-sized hexagonal plates

Curved crystals of mirabilite – like gypsum, a sulphate mineral – hang from the walls of Toronton Cave in Kenya

quently Mammoth Cave contains some of the most extensive gypsum deposits known in any cave. Gypsum very rarely forms stalactites; instead it forms crystals, even on the cave ceiling. And these crystals grow from their bases, so that they are virtually extruded from the fissures which provide the water supply. If the crystals grow faster on one side than the other they develop a curved form, and a group of curved crystals splayed out from a point is known as a gypsum flower. There are thousands of almost pure white flowers, some up to eight inches across, covering huge areas of the walls of Mammoth Cave. Gypsum deposits may also grow, again by alteration of iron sulphides, in the clay in caves. But these are not curved crystals; instead they are thin, transparent but lustrous blades, and they can sometimes be seen sparkling in the cavers' light, where they sit on the tops of clay banks. Gypsum Cavern in the Easegill caves of England gained its name from the abundance of these crystals on its floor.

The term 'ice cave' is commonly applied to two completely different types of structure. First there are the water-eroded caves which are visible at the snouts of many of the world's glaciers. Formed by sub-glacial meltwater, they may be large and impressive but are rarely long or even very stable – they change form from year to year. In complete contrast is the second type – limestone caves which are partly filled by ice. Naturally, these ice deposits occur only in the colder parts of the world – in the caves of northern Norway, the American Rockies and the higher parts of the European Alps, as well as in the lava caves of Iceland.

Caves with two entrances at different levels develop air currents through them. Deep within the rock they tend to maintain almost constant temperatures, so that in winter their light, relatively warm air draughts upwards and in summer their heavy cool air drains downwards. So if the outside temperature conditions are right a ventilated cave can stay frozen all year and all the drip water that enters it freezes into ice. It has also been suggested that some cave ice is left over from the Ice Ages and has not yet had the chance to melt away. And the fact that cool air is heavier than warm air can account for the ice in badly ventilated caves with only one, high entrance – they just act as cold air traps. Two of the most fabulous ice caves in the world (and both

now show-caves) are the Dachstein and Eisriesenwelt caves, only a few miles apart in the Salzburg Alps of western Austria. Each has a maze of large passages, with the sections near their lower entrances liberally decorated with ice. Some of it is opaque white, some almost transparent blue, and it forms every shape of massive stalactites, curtains and stalagmites; down the main passage of the Eisriesenwelt runs a veritable subterranean glacier.

In complete contrast to the massive splendour of the Austrian ice cave formations are incredibly delicate ice crystals formed on some cave walls by condensation of water vapour in the air at below freezing temperatures. A fabulous display of these is found in the Plateau Mountain Ice Cave, high in the Canadian Rockies. From its lonely entrance, the walking-size passages slope gently downwards, and they only extend a few hundred metres. But about half of them are completely lined by ice crystals. The limestone walls are invisible through this thick layer of ice – mostly in hexagonal plates up to eight inches across. And the crystal faces reflect light, so that the whole cave sparkles as anyone moves down it with a lamp – it's the nearest thing imaginable to the proverbial fairyland.

A translucent tower of ice links ceiling to floor where dripping water has entered the frozen Grotte Casteret high in the French Pyrénées (far left). In the Victgeliier Cave in Iceland (left), ice stalagmites break the monotony of the black lava tube

7 Life i

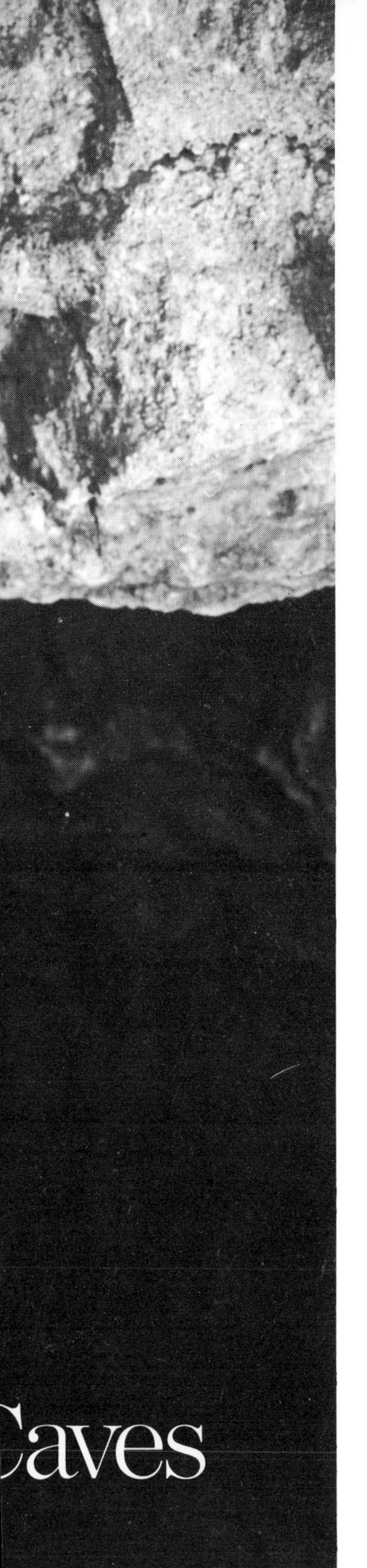

Caves

To any form of life, caves represent a most unusual habitat; indeed, they provide one of the most distinctive natural environments known. Consequently any form of animal or plant life that can exist in caves is itself very distinctive – it is normally specialized to a degree which quite clearly separates it from its surface-living relatives. Nearly all the peculiar features of the cave environment are hostile to life (darkness is the obvious example) but one feature – the almost constant temperature – can be favourable to life. In summer, caves are generally colder than the outside world, but in winter most are relatively warm. It is this one small good point about life in caves that is largely responsible for the surprisingly broad range of animals which live all or part of their lives underground.

Inside a cave, beyond its entrance zone, it is not just dark, it is totally black. There is no light at all, and this affects nearly all forms of life. Nearly all plants depend upon light – it is their main energy source, in that they use photosynthesis to produce the sugars which are their basic food. No light therefore means virtually no plants, except for a few types of fungi which do not photosynthesize. And as so many animals depend in turn on plants for their own food supply, this places a marked restriction on the range of animal life in caves. The darkness is probably the most far-reaching and influential of all of the peculiar characteristics of the cave environment.

The total lack of light means that even the strongest of eyes are completely useless. So some permanent inhabitants of caves do not have any eyes at all, and even temporary cave dwellers have evolved with great improvements to any other sensory powers which they may have. Another means of overcoming the lack of natural light is the construction of artificial light. But only two cave dwellers have done this. One is man – he takes bunches of burning twigs, or electric lights, into the caves he visits; but man is a unique animal in many other ways too, and it is not really surprising that he has included the darkness of the caves in the list of environments which he has conquered. Far more remarkable is the second creature to light up caves – the glow-worm. There are quite a number of animals which produce their own light, including both the glow-worm and the firefly, but only one is known to inhabit caves. The Waitomo Cave in New Zealand is famous for its unique 'glow-worm grottoes' – a series of chambers along an underground river with their roofs covered with spots of light from thousands of 'glow-worms'. The Waitomo 'glow-worm' is in fact the larval stage of a fly, of a type known nowhere in the world outside Australia and New Zealand. However these incredible little animals do not generate their lights to see by; in fact it is not known just why they do light up. Some types of bioluminescent, surface-dwelling animals develop their

Found deep in a Venezuelan Cave, this cricket has no eyes at all

light for the purpose of attracting mates, but this would not seem to apply to the Waitomo glow-worms – since they are most strongly luminous in their larval stages.

The cave environment is popularly thought of as 'cold, dark and damp' – not an unreasonable generalization but, except for the darkness, not altogether true. A cave's temperature remains constant right round the year – at a figure very close to the annual average temperature outside at that particular locality. Most people probably visit a cave on a summer afternoon – when it is relatively cold – but the same cave visited during a winter evening will feel very warm indeed. So caves attract a certain range of animals who simply take advantage of this winter warmth, while the permanent cave dwellers have become accustomed to constant conditions; some permanent inhabitants may even die if taken from their cave to the outside environment, with its relatively harsh daily temperature changes. Most caves have some water in them and will therefore contain humid atmospheres in contact with it. In a cold cave this merely adds to the overall damp discomfort of the place. But probably ever more unpleasant to man is the combination of high humidity and high temperature in some tropical caves. These can have a really oppressive atmosphere, which is however conducive to colonization by an even more active range of animal life, of the types that thrive in such hot, moist conditions. In complete contrast there are some caves which have no water at all, and consequently have very dry atmospheres – such a cave can provide a very comfortable haven during a cold winter.

Food is rarely in abundance in the cave environment. Every form of life can exist only if it has a suitable food-supply. Outside man's synthetic modifications, animals obtain their food either by eating other animals or plants. Every natural environment has its own 'food-chain', where each life form takes its food supply from the next form down the chain. And at the bottom end of nearly every chain is plant life – which obtains its energy directly from the sun, by photosynthesis. The caves have no sun, no energy, no plants and no food of their own. So life can only exist in caves when it is based on food brought in from outside. Either the animal ventures outside, or food is carried in from the outside for it. The permanent cave dwellers obtain their food only in the second method, and not surprisingly they are in the minority among the range of animals who do live at least some of the time in caves.

Permanent cave dwellers are known as troglobites. They are the most peculiar and the most specialized of the cave animals, and they just form one group out of the three into which all cave life can be divided; the other groups are the troglophiles and the trogloxenes. The troglophiles also live their entire lives inside caves, but they differ from the troglobites in that they are also found living outside caves; not surprisingly they thrive in the entrance zones. In contrast, the trogloxenes are dependent on the daylight environment; they are just visitors to the caves, though regular visitors, returning to the caves either every day or every year, but mainly feeding outside the caves.

Electric storage heaters are a relatively new invention of man. They consist of piles of bricks which are heated up so that the warmth will slowly spread from them over a long period of time. But the trogloxenes discovered this principal millions of years ago. Every summer the rocks of the world are slightly heated up, and the warmth which is dissipated from them during the winter is most concentrated in the caves. Among the largest of all known bears was the cave-bear of northern Europe – a trogloxene who died out at the end of the Ice Ages. He was an impressive animal, standing about ten feet long and about half as high, but perhaps fortunately for our ancestors he was essentially herbivorous. Every winter he retired into a convenient cave to hibernate and await the warmer weather. The Bärenhöhle in the Schwabian Alb of southern Germany must have been a popular lair, for it now contains hundreds of bones and skulls all over the floors of its seven linked chambers. This is not to suggest that all the bears died in the caves, but just that the majority of well pre-

Food chains in the cave environment show the close interdependence of all the life forms

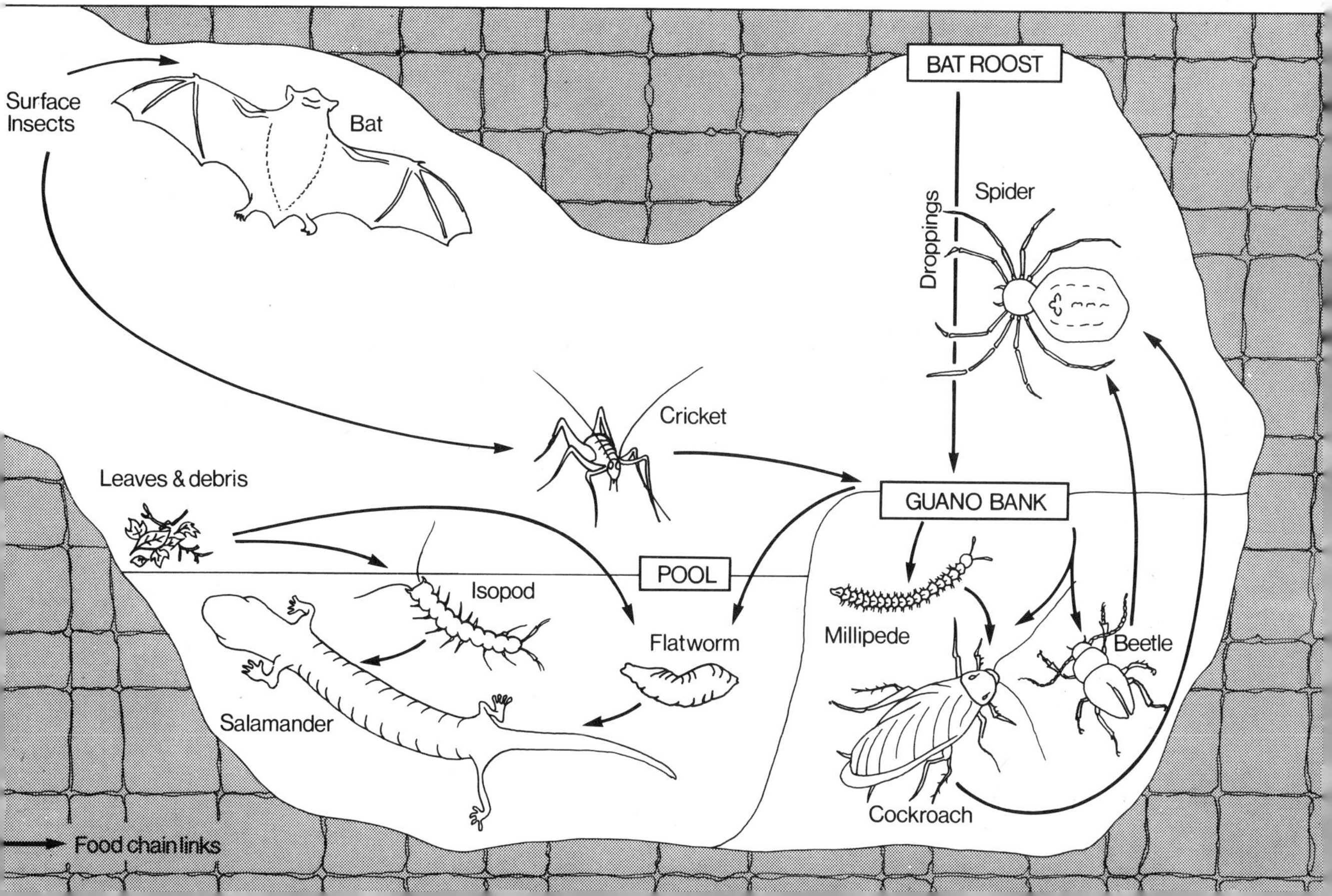

served skeletal remains are those isolated from decay in the old cave-lairs.

Bones of many other animals have been found in caves throughout the world. In northern Europe these include tigers, rhinoceroses and mammoths among a huge collection of Ice Age animals. Some of the animals were trogloxenes, but others just fell into and died in caves, and others had their corpses dragged into the caves by the trogloxenes. One of the commonest, and one still active today in parts of Africa, is the hyena. Hyenas are scavengers, so they will frequently drag an animal limb into the dark zone of a cave, where they can devour it in comfort. They also sleep in caves, but for this use they generally select small and constricted openings, whereas any convenient cave will do for their periodic feasts. They still follow this life style in some of the lava caves of Kenya, where a whole range of other large animals occasionally shelter in cave entrances. The comparison must be made between these African caves today and the caves of northern Europe during the Ice Ages for the latter contain many bones of large animals, with the hyena often being the commonest represented. East Yorkshire's Kirkdale Cave, West Yorkshire's Victoria Cave, East Derbyshire's Creswell Caves and Somerset's Hyena Den are just some of Britain's caves where the abundance of bones in the sediments indicates their popularity with the Ice Age trogloxenes.

At the opposite end of the size range, there are two trogloxenes which are particularly common in the caves of North America. The harvestmen – or daddy-long-legs – regularly hibernate through the winter, clustered in their thousands on cave walls and ceilings; each summer they flock to the outside for a season of activity. In contrast the other type is a daily visitor: cave crickets spend their days in the caves, but migrate outwards at dusk to spend the night searching for food. Both are regular cave visitors – typical trogloxenes.

There are two types of birds which live far enough into caves to be counted as trogloxenes. One is the famous oilbird, *Steatornis caripensis,* of South America. Known locally as the guacharo, this bird was 'discovered' by the German naturalist Baron von Humboldt in 1799, in the Guacharo Caves near Caripe in Venezuela. It has since been found elsewhere in Venezuela, as well as in caves in Trinidad, and in the Lechuza Cave in the Amazon forests of Peru. With a body length of about eighteen inches and a wingspan which may be as much as a yard, these brown owl-like birds make an impressive spectacle as they fly through a totally dark cave. They nest and breed in the caves, but fly out each night to feed on seeds and fruit which grow in daylight. Anyone visiting a cave with oilbirds (so named because local people used to take the blubber from young birds for oil) is made

A whole variety of animals' bones adorn a hyena's lair in the Kajaido Cave in Kenya

acutely aware of their presence – by the infernal shrieking noise they make when disturbed. Unfortunately this tends to mask the incessant clicking noise which is their navigation system – for they fly around in complete darkness by using their own natural, built-in echo-location device. But unlike man's artificial equivalent, radar, the frequency range the oilbirds use is audible to the human being, as a sequence of low frequency clicks. They do also have eyes and in daylight they fly quietly, as their echo-location apparatus is superfluous.

The double quality, of eyes and an echo-location device, is also possessed by the other type of cave bird, the salanganes or *Collocalia fuciphaga*. These live in the caves of the East Indies and eastern Asia mainland, where they are famous for their nests, from which the delicacy 'bird's nest soup' is made. They feed outside the caves in daylight but sleep in nests which are commonly stuck to the cave walls 150 feet or more above the floor of some of the massive cave passages which abound in that part of the world. More than a mile of passage in the Niah Great Cave of Sarawak averages 300 feet wide and 200 feet high, and houses thousands of the salangane birds. Their nests are held together, and attached to the cave walls, by a sticky secretion which the birds produce, and it is this that is in demand for the famous soup.

All the trogloxenes make an invaluable contribution to the pattern of cave life, in that they bring in food from the outside. They can then provide food which can support a second assemblage of cave animals, including the troglobites who never venture outside but are reliant on food carried in. This contribution to the food chain may be in the form of dead, or even live, individuals; but far more important are the droppings. Guano is usually a phosphate-rich sediment consisting largely of animal droppings, and accumulates rapidly in caves where it is not subject to erosion, therefore especially in dry abandoned caves. The salanganes and the guacharos are both prolific producers of guano, and large populations of cave crickets in some American caves have produced enough guano to support further types of animals. The cave food chain commonly starts with the trogloxenes' droppings.

Besides the birds and bears and crickets, there are two more very important trogloxenes. One is man. His use of caves is many and varied but he rarely contributes significantly to the food chain. An exception is in Germany's Bärenhöhle where a pile of human bones has been found, perhaps originally placed there after a spread of a fatal plague. In parts of the Celebes today, the natives bury their dead in sand deposits quite considerable distances inside caves – whence the corpses must provide a welcome feast for the cave-dwelling scavengers.

Horseshoe bats in a cave near Buckfastleigh in England have their squeaking recorded by a caver with a special portable receiver

Last and most remarkable of the trogloxenes are the bats. Bats are small furry flying mammals, whose wings are formed of membranes linking their fore-arms to their bodies. They occur all around the world and come in a considerable range of shapes and sizes. Among the smallest are the Vespertilionids, which have rather mouse-like faces, and bodies normally about two inches long – though their wingspan is usually around nine inches. This group includes the pipistrel, so common in America, as well as all the species of *Myotis*, including Britain's Daubenton's bat and Natterer's bat. Commonest in England are the slightly larger horseshoe bats, which are readily recognized by their horseshoe-shaped nasal membrane. However the greatest numbers in America are probably of the free-tailed bat – this is the type found in colonies of a million or more in the caves of the south-west, including Carlsbad Caverns.

Some of the less abundant bats are probably rather better known than these very common forms. Pride of place must go to the infamous vampire bat. Not really the terror that legend makes him out to be, the vampire bat has a body length seldom greater than three inches and only lives in the tropical areas of Central and South America. A flattened nose and razor-sharp front teeth enable him to make a tiny cut in the skin and drink the blood of sleeping warm-blooded animals; but he takes so little blood that he offers a far greater danger to his 'victim' by transmitting rabies to them!

The two larger groups of bats are the fox bats and the fruit bats. The former are particularly common in the caves of eastern Asia, and their wingspan commonly exceeds twenty inches. But even these are small compared to the fruit bats, whose wings may span nearly six feet. It can be very disturbing to meet one of these in a relatively small cave passage, and it is perhaps fortunate that they live on fruit and do not have the same feeding habits as the vampires.

All the bats like cold dark places, where they can either sleep through the days or hibernate through the winters. Roofs of old buildings are popular bat roosts, but equally suitable, and in some cases far more spacious, are caves. But bats are only trogloxenes, for they always feed outside caves. As their name implies the fruit bats are not carnivorous; but in this respect they are the exception. They are also the only bats which rely on their very good eyes for location – they cannot fly in total darkness; indeed they will commonly sleep out in broad daylight, high in trees, and rarely venture beyond the entrance zones of caves.

However, the majority of bats are nocturnal hunters. They fly outside at night and feed on insects, which they usually catch in flight. They are in fact remarkably adept at locating and catch-

Bat colonies

ing fast-flying insects, which indicates the efficiency of their location devices. Like the cave-dwelling birds, they have relatively poor eyes, but exceptionally well developed echo-location systems. The sounds they emit for the echo-location are mainly just outside the range of the human ear; so instead of the clicking and raucous screeching provided by a disturbed guacharo bird colony, a disturbed bat roost just makes a very high pitched, rather ethereal, whistling noise. This quality means that bats too can fly in total darkness and penetrate considerable distances into cave systems. It also means that most stories about the horrors of bats becoming entangled in people's hair are rather exaggerated. However bats can make mistakes, especially when frightened. A caver entering a small crawlway passage leading to a chamber with a colony of bats can strike terror into the harmless little creatures. Their natural instinct is then to fly away, out through the passage occupied by the intruder, and as they all crowd through it is common for them to crash into the body and face of the person. This is a rather unusual caving hazard, but one that is really more frightening than dangerous.

There is doubt as to how bats find their way through complex cave systems – is it by memory or exploration? The greatest movements of bats are always at dusk when they set out on their night's feeding flights. When an entire colony emerges from a cave in a short period of time, the resultant bat flight is one of nature's great spectacles. Perhaps the most famous is that from America's Carlsbad Caverns, when the bats stream out to feed in the nearby Pecos Valley. Normally about half a million free-tailed bats emerge in less than half an hour. They fly up the massive entrance passage in great spiralling sweeps, as if searching for the point where they can break out from the spiral and fly off into the open air. This is in complete contrast to the nightly bat flight from Nepal's Harpan River Cave. There about ten thousand fox bats emerge each evening, but they don't spiral out of the cave. Instead they fly in a straight line along the entrance passage till they approach a wall; they then turn obliquely away until they approach the opposite wall and repeat the process. It is just as if they have memorized the zig-zag route out.

Just before dawn most bats return to their caves to sleep through the day. They sleep by hanging from the cave roof or overhanging walls. Their claws can grip on minutely small projections of the rock; they then fold their wings to their sides and hang upside down, with their faces just visible between their wings. Some bats sleep alone in an isolated corner, while others cluster in great colonies – thousands can cram close together and cover the roof of a cave chamber with a living rug. From below such a colony looks just a mass of fur, with little faces peering out.

A sleeping Greater Horseshoe bat (far left) clings to tiny protrusions on the wall of a cave in the English Mendips. Further south in a Devon cave a whole colony of the same type of bat hangs sleeping from the roof (above), while at dusk each evening the sky above Carlsbad Caverns in the USA is clouded by millions of bats leaving for their nocturnal search for food (left)

When viewing these colonies one is however advised to look, not straight up at them, but obliquely from one side. This is because of the material falling from the colony – almost as a steady rain from densely packed roosts. The falling debris consists of all sorts of indigestible insect remains, together with dead bats, and most prolific of all, their droppings. These accumulate on the cave floor to form great banks of guano; and these guano banks, whether from bats or the cave birds, are commonly seething and heaving with numerous other small animals. Most important among these are the various troglophiles.

The importance of this guano is as food. Its own ultimate source is the insects and plants outside; but it is carried into the caves in the digestive systems of the birds and bats. Together with stream-washed debris and dropped food scraps, it is the vital food which supports the forms of life that do not venture outside the caves. Typical of the guano-dwellers are cockroaches. These scavenging crawlers, up to four inches long and with long feelers, will eat almost anything; in the rather restrictive environment of the caves they have little choice, and it is the non-selective scavengers who are the most successful troglophiles. Cockroaches will eat the guano itself, as well as any animal or vegetable remains in it, and it is quite terrifying to see the speed at which they will eat one of their own kind that happens to die. Cockroaches are rarely found deep inside cave systems – they stay near the entrance where their food supply is most abundant. They are also absent from English or north European caves – it is the large tropical caves which may contain tens of thousands of individuals in a large guano heap.

The skeleton of a small bat has been picked clean by scavengers living on the guano floor of Kimalia Cave in Kenya (far left); in the same cave, scavenging Tenebrionid beetles feed on the bat carcasses (left)

Cockroaches are not the only scavengers of the guano – they are normally joined, in the tropical caves especially by others such as millipedes and beetles. And all of these are themselves food for the predators, of which the most successful are the spiders. There seems to be something about any spider which makes man slightly scared; a huge tropical spider running over the walls of a dark cave is surely even more terrifying. A cave-dwelling spider cannot always copy his surface-living cousins by building a web and waiting for insects to fly into it. There are too few flying insects in many caves, and he must be prepared to hunt and capture the various crawling creatures. Many of the sleek black, spindly-legged spiders of the South American caves have huge leg-spans – commonly over one and a half feet; some of these monsters are even large enough to prey on young bats!

All in all, these caves of the American tropics present a rather repulsive spectacle to the visitor. Thousands of bats or oilbirds flying about in confusion, whistling and screeching, create an impressive sound effect. Then the sense of smell is stimulated as feet disturb the floor of guano and release its ghastly stench. Finally the eyes become accustomed to the light and settle on the seething mass of fist-sized cockroaches and pencil-thick millipedes in the guano, while giant spiders scurry up the walls. Footprints in the guano, left by the visitor on his way into the cave, are gone when he returns, even if after only a few minutes; the floor is a seething mass of life and a footprint soon blurs into indistinction. Most visits to the tropical guano caves are short!

The Harpan River Cave in Nepal's tropical lowlands has yet another surprise for its visitors. The normally used entrance

The caves of Northern and Central America are noted for their blind fish (right) while the fearsome spider (opposite) is an inhabitant of the caves in the Venezuelan jungle – it has a leg span of four inches

emits a gentle but steady draught, which must be cooling and rather restful for the spiders who live there, for they congregate on the walls just beyond the limit of daylight. These spiders are not so large – their legs span about four to six inches – but their grey furry bodies and sharp red eyes give them a fearsome appearance. And as this cave passage is less than a yard in diameter, crawling through it is a nerve-jangling experience.

There is yet another form of life which depends for food directly on the trogloxenes and troglophiles. This is fungus. It is almost the only plant life that can live in the caves, as it does not rely on photosynthesis for food production. Instead it feeds on remains of animals, and then in turn is eaten by others. Along with bacteria, it is a link in the food chain, whereby outside organic material is made palatable to the cave dwellers. The famous 'glow-worm' larvae of Waitomo feed largely on fungus. Any show-cave visitor may notice various plants growing around the show-cave lights – they are unremarkable because they are green, as they are 'fed' by the artificial light. But the true cave plants, the fungi, don't have this light, and unlike many of the cave-dwelling animals they cannot periodically move outside for a 'dose of sunlight'. So they are white. And this lack of colour is also just one remarkable characteristic of the most remarkable animals of all – the troglobites, the true permanent cave dwellers.

Rumours of dragons in the caves of northern Jugoslavia first

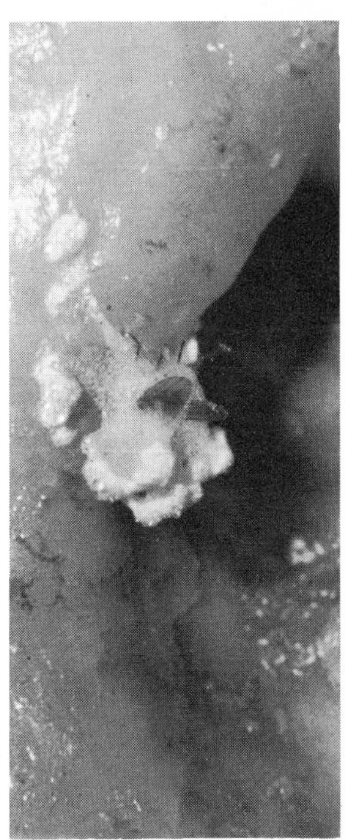

On the stalagmite wall of an English cave a fungus colony feeds on and slowly grows over a single dead fly

appeared in print in 1689, and it was only in the next century that the mysterious cave dweller was recognized as a salamander, and was named *Proteus*. With a length of no more than a foot, it barely matches the mythical ideas of the size of dragons, but it is by far the largest true cave dweller found in Europe. And its ghostly white colour must only enhance its air of mystery. Indeed it is a mysterious animal – in fact a typical troglobite. It spends its entire life in rivers in the many caves of the Postojna area in northern Jugoslavia (but close relatives are also known from as far away as Mammoth Cave in the U.S.A.). The lack of light eliminates its need for colour pigment to protect it from the sun's rays, so it is pure white, that is, except where it is red due to its blood being concentrated near its skin, at the gills for example. Also it has no eyes. It feeds in the only way it can, by scavenging, on any sorts of debris which are washed into the cave by the river in which it lives. But it has replaced its eyes with some form of vibration sensors, for it also preys on other minute forms of aquatic life, the isopods and flatworms which in turn have fed on surface-derived debris. Indeed *Proteus* is a magnificent example of an animal living in an essentially foodless environment: it can live on the minute amounts of organic debris in clays and silts – young specimens have been known to thrive for a year on nothing more than clay. And if all else fails *Proteus* can even survive for periods of years without any food

Bats that live in caves

A Leisler's Bat flies safely through the complete darkness of its cave home (below), and in a Venezuelan cave the bats successfully avoid an intruding caver (far right). A fox bat (right) caught in a Nepalese cave proved to be rabid, and also had living in its fur tiny parasites (centre right) which were malaria-carriers!

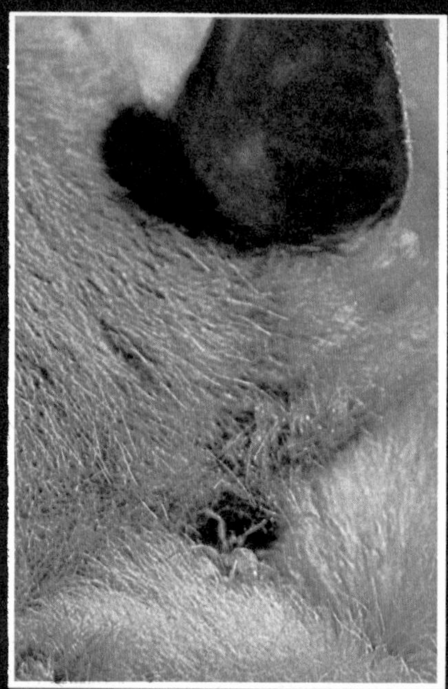

The blind white Proteus in the Jugoslavian cave stream (above left), and the Guacharo bird in the Venezuelan cave (above) both live safely in the cave environment. But the Dall sheep all died when they wandered into the Grotte Valerie in northern Canada (left)

at all! Its incredible feeding – and starving – habits are matched by its reproduction habits; normally *Proteus* produces eggs, but under exceptional conditions it has been known to completely change, and produce live young.

It is almost symbolic that the remarkable *Proteus* should only be found in the classical karst caves of Jugoslavia. But the rest of the world is not without its troglobites. In the U.S.A., the caves of both the Ozarks and Kentucky contain white salamanders very similar to *Proteus*, and while some have small eyes, others are eyeless and rely on vibration sensors to find their way around. The American caves however do have most of the world's cave fish. The Mexican blindfish is the most important, but close relatives are also found in the caves of Cuba and the eastern United States. Different individuals and groups living in separate caves have developed different characteristics, and while some do have eyes it is only those who live in true troglobitic conditions which are totally blind.

Britain's caves are, and always have been, just that little bit colder than those in southern Europe and southern U.S.A., so their entire life systems are rather less developed, and they contain none of the larger more spectacular troglobites. Flatworms and insects less than half an inch long are all that are to be found in most British caves – with one exception. This is *Niphargus* the cave shrimp. Usually half to three-quarters of an inch long, in many a cave he may be found in shallow pools, where there is a bit of fungus or stream-washed debris to feed on. One of the remoter places he has been observed is in the Hensler's Passage in Yorkshire's Gaping Gill cave system. This passage is a crawlway, about 400 yards long and all well under a yard high, though it is amply wide. About midway along it, a shallow puddle no more than three feet across has been known to contain a *Niphargus*. His body is completely transparent and the only colour in him is the red of his blood. So his whole blood system is visible as he swims around in his own little world – no more than a yard across. This lonely troglobite is a strange companion for the passing caver on his long journey through Hensler's Passage.

As long as the cave-dwelling predators have not been too successful, *Niphargus, Proteus* and the other troglobites are almost the last link of the food-chain in the cave habitat. Their food supply is tenuous and their environment is basically hostile, but they do have one advantage – they live in places safe from any other predators; indeed this is the attraction of caves to them. Yet in the food-scarce environment of the caves, even they make their contribution. For when they die, they are the immediate prey for bacteria and fungus, and feeding on them will be yet another generation of cave-dwellers living through their endless night.

8 Man in Caves

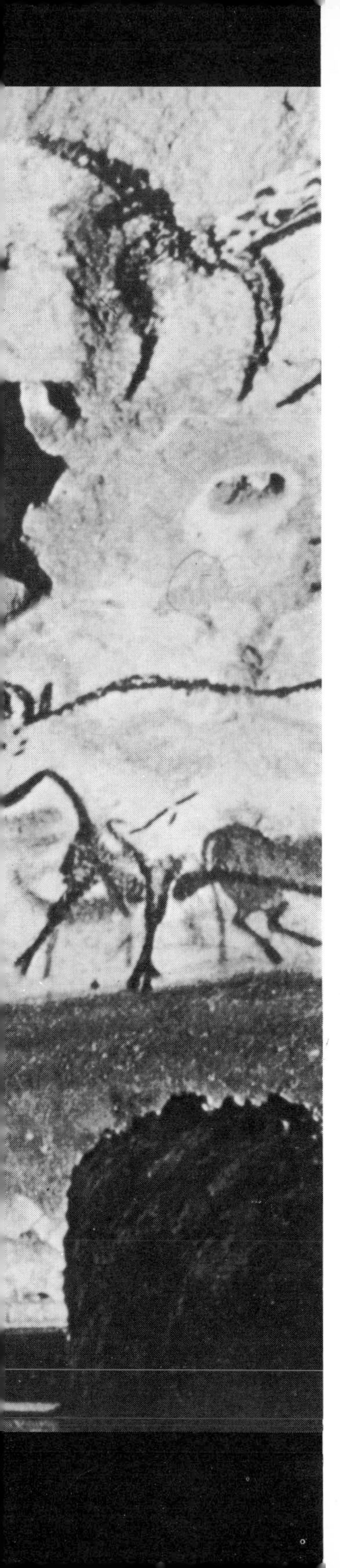

When the Tasaday people were discovered in the jungle of the Philippine Islands, it was no surprise that they were living in caves. The Tasadays are the most primitive people known to exist in the twentieth century world. Only a handful of them are known (they were not 'found' by modern man until June 1971), living in a totally isolated community in the centre of Mindanao Island. They are straight out of the Stone Age, living on the fruits and animals of the jungle, and like so many Stone Age people they retire to rest in caves.

The Tasaday's caves are not great cave systems like those which attract the sporting caver. For such systems would have draughts and perhaps streams flowing in or out of them. Instead they are small hollows in cliff faces, best described as rock shelters, for it is these which have always provided the best homes for primitive man – both Tasadays today and the Europeans of ten thousand years ago. Caves and rock shelters made ideal homesites for early man. Careful selection could usually find one that was completely dry and had a reasonably level floor, perhaps covered by a layer of soft sand. Side openings and wall niches made convenient stores and even separate rooms. Enough daylight came in through the entrance to remove shadows from all but the deepest recesses, and perhaps most of all it was a haven of safety – with the inhabitants completely closed in on three sides. A fire in the entrance at night made it safe from nocturnal predators such as wolves and bears. And finally it was warm in winter and cool in summer.

Cave homesites of ancient man are known from nearly all over the world, but they were concentrated in a particular climatic zone which included most of southern Europe. This is because in the warmer climates man did not need the shelter offered by caves, and the colder climates were too harsh for him in or out of the caves. The limestone areas of southern France provided an ideal environment for ancient man, with their equable climate and numerous caves and rock shelters. It has been suggested that the Dordogne region, east of Bordeaux, was the most densely populated part of the world 20,000 years ago.

Britain was really a little too cold for early man, and it never had a population to match southern France. But it does have a number of cave shelters, and many were inhabited. Creswell Crags on the Derbyshire-Nottinghamshire border provided ideal homesites and were occupied for long periods between the Ice Ages of the Pleistocene. In an area of relatively low relief, which would have been partly covered by forest in the interglacial periods, a low hill of dolomitic limestone is cut by the gorge of a small river flowing eastwards. Steep cliffs, 100–150 feet high, line each side of the gorge for half a mile or so, and the river meanders along a flat floor about a hundred yards wide.

Creswell Crags (above right) in the Midlands of England are broken by a series of rock shelters which were inhabited by ancient man; from the caves on one side of the valley they could look across to those in the opposite cliff (above left)

The Lascaux Cave (previous page) is famous not as a homesite but for its fabulous paintings

Each cliff is broken by a series of rock shelters and small caves, few extending in for more than about fifty feet. They made fine homes. Surrounded by rich countryside, adjacent to a water supply, and easily accessible yet safe and secluded, they even had dry and sandy floors. Though the north bank shelters caught the day's sun, they were not the most popular as residences. The alternation of the day's warmth from the sun with the nightly freezing initiated 'freeze-thaw action' – one of the most effective natural means of rock breakdown – so that the cliffs above the cave entrances were dangerously unstable. The south bank shelters lacked some daily warmth but were at least more safe. The Creswell caves were not modified by man at all – they were just simple shelters.

An Indian village is built in the shelter of the wide sandstone cave in the cliffs of Mesa Verde, USA

This is in complete contrast to the way some of the caves in the southwestern U.S.A. have been used by the Navajo and Pueblo Indians. Canyon de Chelley and Mesa Verde are just two of the many locations where shallow rock shelters have been carved by wind erosion out of vertical red sandstone cliffs. From around 2000 years ago these were occupied by the Indians, who not only lived in them but built houses in them. Some contain complete villages of stone-built dwellings, for these rock shelters were not deep enough or sufficiently well enclosed to make a perfectly natural homesite – they merely served as attractive sites for their villages. Some of the Indian houses were even carved out of the relatively soft sandstone forming the walls of the caves.

This carving of dwellings out of the solid rock has been practised in many parts of the world. The amazing rock houses in the volcanic ash towers of Goreme in western Turkey and the networks of tunnels in the sandstone beneath the town centre of Nottingham in England are but two examples. But like so many others they are completely artificial. Even though Nottingham's tunnels are always known as 'the caves', they cannot be regarded as caves in the true sense.

Though these artificial caves do present a fascinating aspect of the way man lives, the natural caves have an extra special importance in that they were occupied by man much earlier – before he had learned how to carve his home from solid rock. With a few spectacular exceptions, notably Kenya's Olduvai Gorge, it is caves which have yielded the best-preserved remains from which the archaeologists of the world have deduced the prehistory of our distant ancestors. The sediments on the floor of the caves are the archaeologists' hunting ground. Over the thousands of years wind-borne dust accumulates, rock fragments fall from the cave ceiling, and animals, including man, dump their rubbish; all this combines to build up the cave sediments. Not only does this therefore bury and preserve the older remains, but it means that the material accumulates in sequence, with the youngest at the top, and the archaeologist can almost literally scrape back the successive layers of history as he digs deeper into the cave floor.

The sequence therefore recorded in the cave deposits can be most instructive. It reveals for example that England's Creswell Caves were occupied by man for a number of isolated periods of time. In between his visits, the same caves and rock shelters were used as hyena lairs. This alternation of residents can then be related to environmental changes – in this case to the climatic fluctuations through the sequence of Pleistocene Ice Ages. Furthermore, cave sediment sequences can reveal the evolution of the different types of ancient man, by study of the skeletal remains, and also the evolution of his implements. Axe-heads and other tools, chipped or carved from rock or animal bone, preserve very well, and once their changes in pattern and style have been studied in a well calibrated sequence of sediments, they can be used to date isolated discoveries in lesser caves and sediments.

Not surprisingly, China's immense limestone regions have yielded evidence of cave dwellers. The caves of Kwangsi in the southwest of the country have yielded numerous remains and stone implements, but little conclusive evidence has been found of inhabitants of this region before 10,000 years ago. Instead, quite a small fissure cave in an outcrop of limestone about twenty-five miles southwest of Peking has revealed China's most spectacular remains. The Chou-k'ou-tien Cave was excavated

The rock houses of Goreme in Central Turkey are not in naturally formed caves; they were excavated out of the soft rock by the early settlers in the region

in 1921 and in it were discovered skull fragments of a man who lived 500,000 years ago. Officially named *Sinanthropus pekinensis*, and better known as 'Peking Man', he appears to have been one of the first men to live on earth.

America has no human remains whose ages can match those of China and the Old World. The Stone Age Americans were thriving at the same time as some of the Mediterranean lands were reaching the peaks of their civilisation. Around the year 1200 B.C., naked warriors made their home in Russel Cave, in the limestone hills of Alabama, at the same time that Moses made his exit from Egypt. With an opening thirty feet high and three times as wide, Russel Cave had then already been occupied for over 6,000 years; this long period of activity has left a magnificent record in the sand and clay of the modern cave floor. Skeletons, pottery, fish-hooks, bone lamps, axe-heads and fire-hearths have all been excavated to give a very complete picture of the life style of these early Americans.

Russel Cave is an obvious opening – a site which could barely be missed, by either the Stone Age home-maker or the modern archaeologist. But some of today's discoveries of ancient remains are found in the most remote places. Cavers who broke into Salts Cave in Kentucky thought they had entered it for the first time ever; but they found footprints in the soft mud, left behind by an Indian, probably searching for minerals – some 2,400 years earlier. Even more remarkable was the discovery made by explorers in Stoke Lane Cave in the English Mendips. They crawled down 300 yards of low wet passage, dived through a completely submerged tunnel, traversed high in the roof of a great chamber and only then entered a boulder-strewn cavern. And in the middle was a fire-hearth with charred wood. Primitive man had been before, but surely not by the same route! The cave was mapped, and the chamber was found to lie just below a collapse depression in the woods above the cave. That chamber had originally been open to the outside, and it appears that some passing Stone Age hunters had just wandered in for a rest and had built a small fire just in the range of daylight. But of course, like men, caves change.

It is always the remarkable and the typical which make the most interesting discoveries. And so it is with caves too; these many fabulous dwellings of ancient man are probably less known to most people than the caves of Khirbet Qumran – or at least their contents. This series of caves lies in the desert of the Judean wilderness on the northwest shores of Jordan's Dead Sea. In 1947 they were found to contain a collection of leather manuscripts and papyri – now commonly known as the Dead Sea Scrolls. These appear to have belonged to a Jewish religious library sited in the area during the third century, and their fortuitous preser-

Excavation of cave sediments by archaeologists

A limestone quarry in Devon, England, broke into a sediment filled cave (above), and before the quarry face bit further into it, local archaeologists excavated and recorded the cave contents. They first cleaned up the face (centre) to record the sediment sequence, and then steadily dug the material away slice by slice (above right)

vation in these lonely caves has given invaluable insight into the very birth of Christianity.

Sediments in the home-site caves contain plenty of bits of pottery and bone implements, and prove a mine of evidence into the way life of the cave-dwellers. But their ornaments and jewellery are more rarely found: the inhabitants didn't leave such valuables lying around – they took more care of them. However, many primitive peoples did intentionally place their valuables in one certain type of cave – that which had some religious significance. And it is the sediments of these, the votive cave sites, which yield great riches to the archaeologist. Victoria Cave, in the English Pennines, is famous for the archaeological remains found in its thick layer of sediments. There is clear evidence that it was a homesite, first for animals and then for man in early times, but during the Romano-British period the accumulation of sediment had blocked up its main entrance. Only a small slot remained open in the roof and this must have been something of a shrine into which offerings were cast before

it was walled up. Consequently when it was rediscovered and entered in 1838 the floor was liberally covered with a whole horde of Roman coins and fragments of metalwork. Later excavation of the older sediments revealed its occupational use but never repeated the treasures dating to the time when it was a votive site. The Blue Holes of Mexico's Yucatan Penninsula have similarly been used as votive sites in almost modern times. The low-lying limestone plains of the Yucatan are broken by many of these round water-filled potholes, and unproven stories abound of the great treasures that may lie in the unplumbed depths of some of them.

The worship of caves is probably associated in most people's minds with the rather more mysterious religions of the East. And indeed though it does not have the monopoly of religious cave sites, the East does have some magnificent examples. It is part of their tradition that the Hindu peoples should wherever possible adopt a cave as a shrine and place of pilgrimage. Caves have an undeniable connection with the Earth, and in Hindu terms this

The famous Dead Sea Scrolls were discovered in the lonely caves of Khirbet Qumran, in the heart of the barren wilderness of Judea (left)

Since this photograph of Victoria Cave in the English Pennines (below) was taken a century ago, more than thirty feet of sediment has been excavated from the cave floor by successive teams of archaeologists

is the source of all things and therefore inextricably tied in with the gods. Furthermore the phallic connotations of many stalagmites make them popular symbols for worship within the sacred caves.

Caves and springs alike are regarded as Hindu shrines in the limestone areas of Indian Kashmir. At the top end of the Vale of Kashmir, five large springs emerge from the limestone mountains; all are regarded as holy, because they are providers of apparently endless pure water. The springs are now trained so that they emerge in pools, surrounded by white marble pathways and overlooked by small temples, with the water flowing away through immaculate ornamental gardens. Even more revered is the nearby Amarnath Cave at an altitude of over 13,000 feet in

the Kashmir Himalaya. Little more than a rock shelter, it is about fifty feet high, wide and long, opening in the cliff of a rugged mountain valley. The one ice stalagmite inside it is the particular object of worship, and is visited by thousands of devotees each year, mostly during an annual June pilgrimage. Unfortunately the cave is two days hard walk from the nearest road, across barren mountain passes periodically swept by blizzards. Furthermore the claimed healing properties of Amarnath attract many ill, and old, pilgrims; the death roll on the June pilgrimage frequently exceeds thirty. And one is tempted to wonder if those that survive the rigours of the pilgrimage, and then consider themselves miraculously cured, were ever less than healthy before they started?

The holy cave of Amarnath opens out over the rugged limestone landscape of the Kashmir Himalaya

Further east along the Himalaya Mountains, the people of Nepal seem to regard caves with one of two extreme views – they will either not go near a cave at all due to fear of the unknown, or, if the cave is holy, they will surmount any obstacle to visit it. Even many of the Sherpas, utterly fearless on the huge mountains of their homeland, will shy away from most caves. But the Gupteswary Cave, not far south of the Annapurna mountain group, is different; it is holy, and has therefore even been visited by the King of Nepal. A hundred yards of narrow, walking-size passage lead to a scramble up into a large chamber, where a single massive stalagmite is the object of worship. The main difficulty in visiting this cave is caused by the fact that it is a Hindu holy place – so that the entire length of wet stony passage must be traversed in the traditional bare feet.

In the Mount Everest region, religious festivals are held in the huge chambers of the Halesi Cave. But this is rather eclipsed by the massive Batu Temple Cave in Malaya. This has been adopted as a shrine by the immigrant Hindu community of Kuala Lumpur, only about six miles away, and is the gathering point for tens of thousands of devotees during the annual Thaipusam Festival. The cave is a single chamber, 150 yards long and a hundred feet high and wide, situated in the isolated Batu Caves Hill. The outer end has a huge open entrance, and there is a large roof skylight at the inner end. Consequently the whole cave is dimly lit by daylight, and resembles a massive, smoky cathedral when crowded with pilgrims all burning their offerings to their gods.

Europe too has its share of religious caves, even without counting the sites revered by stone-age men. The cave of La Balme in southeast France has a chapel built in its entrance, though the cave now has a more commercial use and most of its visitors come for guided tours of the show-cave. In the Cantabrian Mountains of Spain the spectacular resurgence cave of Covadonga is held in considerable local esteem as a shrine. Cavers have been exploring, via other entrances, the cave system behind Covadonga, but have always ruled out the possibility of emerging from the resurgence cave should a way through ever be found. But most famous of all European cave shrines must surely be Lourdes, in the French Pyrénées. There in a small cave, it is believed that the Virgin Mary revealed herself to a young peasant girl, Bernadette Soubirous. That was in 1858 and since then the grotto has been regarded as a Roman Catholic shrine which now receives about two million visitors a year – many in search of miraculous cures.

Both modern and ancient man have frequently regarded caves as mysterious objects to be revered or worshipped. In more recent times, structures have been built in caves to accommodate the religious visitors, but the signs of religious activity in

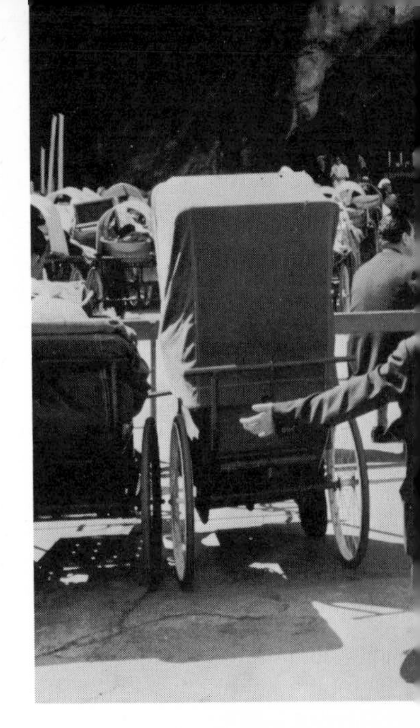

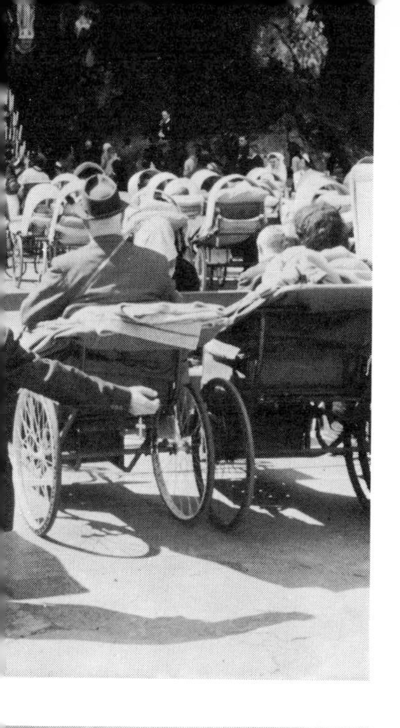

earlier times are often found in various forms of cave art. Sculptures, carvings and paintings, all forms of cave art, have been known and studied for well over a century, and though they provide some of the most spectacular facets of the world of caves, much about them is still clouded with mystery.

Why did men paint, and carve, the walls of his caves? First, if he wanted to paint anywhere he would do so in the caves, because the walls presented suitable surfaces protected from decay and exposure to the weather. And then he did not paint his caves purely for decoration; very little cave art is found in the parts of caves that were inhabited by man – there is rarely more than an odd painting, perhaps serving as an indication of better things in the inner reaches of the cave. The Huagapo Cave in Peru has some fine Inca paintings, at least 10,000 years old, in the entrance chamber in sight of daylight; but this seems exceptional. Southern Europe, France and Spain have the world's best examples of cave art and most are found deep inside the caves. The Niaux cave in the French Pyrénées has a massive main passage

Batu Temple Cave in Malaya has a massive stalactite-draped chamber (left), and its air is heavy with incense burned by the thousands of visiting pilgrims during each Thaipusam Festival (far left). Equally devoted are the hordes of pilgrims who go each year to the cave shrine at Lourdes in the French Pyrénées (above left)

which for nearly a mile is devoid of any art features. Then at a slight widening of the passage one wall is covered with a mass of very beautiful paintings; the artists for some reason ignored all the earlier passage which they must have walked through, on a journey none too easy with the crude lights available to them. In the same region the Trois Frères cave has its best art almost hidden in a small chamber only reached by a narrow and difficult passage. Indeed, this remoteness of so much of the cave art explains why it has only been discovered relatively recently; old entrances had collapsed, and only the more enthusiastic explorations of the last hundred years have rediscovered some of these ancient decorated caves.

It appears that there were perhaps two purposes behind the labours of the stone-age cave artists. Certainly many of the art forms were symbolic, ceremonial and crudely religious. But others seem to have little purpose, and can only be interpreted as a record of man's past environments. These can prove very interesting: perhaps the most remarkable of this type are the cave paintings of Tassili in southern Algeria. There, a number of rock shelters lie in the barren Ahaggar Mountains right in the centre of the Sahara Desert. They contain paintings of men, rich forests and all sorts of animals – the giraffe being the most easily recognizable. The only possible explanation is that the Ahaggar was once a populated, vegetated and well watered area, thousands of years ago before the climate changed and the Sahara Desert expanded to its present size.

Among the oldest forms of cave art are the sculptures – both bas-relief and fully three-dimensional. The Tuc d'Audoubert and Montespan caves, both in the French Pyrénées, are famous for the statues of animals in them. The statues, beautifully detailed, are made out of clay, and represent various animals – bears, lions and bison. Many are riddled with holes where they were stabbed with spears, and one of the bear statues had the head of a real young bear held on its neck by a wooden pin. In front of the statues the clay floor was rhythmically marked from the stamp of bare feet. Clearly the cave was used for ritual ceremonies, about 20,000 years ago, probably when the hunters entered the cave and acted out what they hoped would happen in real life when they went out in search of food.

Etchings and paintings on the cave walls tend to follow the same pattern. Many of the depicted animals are headless, wounded or bearing spears plunged into their flanks. The etchings are the simpler, cruder and probably older forms of cave wall art. Commonly the rock has been scratched and engraved by crude instruments, mainly made of flint which is harder than any limestone. But even simpler are the finger impressions in the soft and friable walls of the Koonalda Cave, beneath Australia's

Nullarbor Plain; and these drawings date back 20,000 years. Cave art reached its finest form with the true paintings. Charcoal and manganese oxides provided the black for the artwork, and a whole range of reds, browns and yellows were obtained from iron ochres. Applied with finger, brush or spray, or even with a simple colour stick, these gave the world's earliest artists all the scope they wanted. And they gave us the amazing painted caves, so many of which have been found in France and Spain.

It is almost beyond comment that the world's two finest painted caves were both discovered by dogs. The cave of Altamira near Santander in northern Spain was the first to come to light. In 1868 a hunter's dog chased a hare into a crevice and became stuck. He was released by his master, who noticed a cool draught from below, and pulled away a few rocks to open his way into the cave. The hunter told the landowner of his discovery, and the latter found some prehistoric implements on the cave floor. Then in 1879, Sautuola, the landowner, returned to dig for more implements, and brought along his twelve-year-old daughter. Fascinated, she wandered further into the cave, and there for the first time saw the fabulous paintings of bisons, horses and boars. Sadly, the paintings were so beautiful that the experts and scholars of the day denounced them as a grand fraud. It was not until the turn of the century that the Altamira cave was commonly recognized as one of the world's finest showcases of prehistoric art.

The second great canine cave discoverer set to work in 1940, on a hillside in the Dordogne area of France. This dog fell clean down his narrow crevice and could be heard barking in the cave below. The five boys out with him dug out the cave opening and lowered themselves down. There by the light of a box of matches they first saw an unbelievable collection of paintings. The cave is now known as Lascaux. Though having no more than a few hundred yards of passage, its walls and ceiling are covered by an unrivalled number of paintings of horses, bulls, bison, lions, rhinoceroses and deer. The colours are red, yellow, orange and black, the artistic quality is of the highest order, and they are 17,000 years old. Isolated in the cave, the paintings had been beautifully preserved. But soon after its discovery Lascaux was opened as a show-cave – only to be closed again in 1967. This was because the passage of thousands of people through the cave had led to an alga growing on and over the paintings. Access had to be very strictly limited. However at this moment a perfect copy of the cave is being built in concrete on an adjacent site. Then it is being painted exactly as the original. It is a far-seeing project, but the only way in which visitors can once again experience the splendour of Lascaux, without forever destroying the handiwork of our artistic ancestors.

9 Caves of the World

Practically every country in the world has some limestone and some caves; happily, however, not all the world's caves are the same. Different types of limestone, different geological structures and histories, and different climates, all combine to give almost surprising contrasts between caves of different regions. It is noticeable that some countries, France for example, have huge numbers of long and deep known cave systems, while others, Nigeria for example, have almost none. At least part of this difference lies in the extent of exploration and the popularity of caving as a leisure occupation. The 'western world' therefore has more than its share of *known* caves, while countries like Iran have huge regions of cavernous limestone, though very few caves have (as yet) been explored. Undoubtedly the story of caves and caving must start and end in Europe; Jugoslavia, France and Britain saw the most important early developments in caving, and today Alpine Europe contains the great majority of the world's really deep known cave systems.

The cave regions of Britain are quite specific. An area covering the Yorkshire Dales from Wharfedale to Ingleborough and overlapping into Lancashire and Westmorland contains the greatest concentration of caves in Britain. That region straddles the Pennines, and further south in the same range of hills, the Peak District of Derbyshire again contains important caves. At the other end of the country, the Mendip Hills in Somerset and the top end of the famous 'Valleys' of South Wales are also great cave regions. Ireland has a far greater proportion of limestone than mainland Britain, and caves abound in Counties Clare and Fermanagh in particular.

Most famous of Britain's caves is Gaping Gill. A stream tumbles down the slopes of Ingleborough and then takes a vertical leap of 360 feet down Gaping Gill Hole. The foot of this shaft opens into a wide flat-floored chamber – first explored by the Frenchman Edouard Martel, in 1895, when he diverted the stream and laddered the shaft. Martel saw no other passages, but subsequent visits by Englishmen revealed many miles of dry tunnels leading off from openings in the chamber walls. Four more tiny streams on the fells above were followed down successions of shafts to link up with the main system of caves. Most remarkable of the many discoveries was made by a caver called Eric Hensler in 1937. Completely alone, he crawled into a low opening near the main chamber. For 200 yards he wormed his way down a passage little more than a foot high, though frequently more than fifteen feet wide; then he crawled through another 200 yards of the same bedding plane passage – this section was easier but still nowhere more than three feet high. Hensler's Passage, as it is now known, is unique in Britain, and that first solo exploration must have required nerves of iron and the

Opposite: The Harpan River cascades down a massive rift in its own cave in the foothills of the Nepal Himalaya

The main passage of Peak Cavern (opposite above) in the English Peak District is a large clean tube with a small stream on its floor. But further north in the Pennines, Easter Grotto (opposite below) in Easegill Caverns is famous for its array of stalagmites and straw stalactites. The long individual straws (left) hang from the rounded roof of the Dan-yr-Ogof cave in South Wales

The decorated roof of the Gavel Pot streamway in the English Pennines almost appears to be supported by the stalagmite pillars resting on the rocky ledge

sort of confidence that only comes with competence. Beyond that crawl Hensler found a magnificent stream passage which ended in a gloomy sump. But cavers do not give in easily, and in 1968 a tight hole over the sump was opened up, and led into another mile of fine passage, aptly named the Far Country.

The protracted exploration of Gaping Gill was partly due to the nature of the cave, in particular the many sections nearly blocked up by the sediment and calcite fills; but it was also due to the relatively inadequate techniques of the early years of caving. The nearby Pippikin Hole provides a complete contrast. In 1970 a short entrance shaft led down a few climbs to an impossibly narrow slot. Explosives were used to make this, and another narrow rift just beyond, passable for slim cavers. The way was then open down a further series of narrow squeezes and a few hundred feet of difficult passage. Beyond lay a magnificent series of caves – four miles long in total. And though they provided every kind of difficulty for the cavers, nearly all were explored in the first four week-ends after the entrance was blasted open. Roaring streamways, long lakes, spectacular cascades, spacious galleries muddy crawls and beautiful stalagmites combine to make Pippikin a classic, both a pleasure and a challenge to visit.

Down at the southern end of the Pennines, the Peak District has relatively few cave systems which provide sport or spectacle to match those in the north. The village of Castleton is well enough known for the number of show-caves around it, and the same area contains the finest of the non-commercialized caves too. Right at the edge of the village a huge gash in the cliff marks the spectacular entrance of Peak Cavern – so large and sheltered that a ropeworks has been standing there since the Middle Ages. But Peak is also one of the most important resurgence caves in the Castleton valley, and the passages which carry the stream out to daylight are large and spectacular. Most of the streamways were only discovered in 1947, when, on the same day, cave divers got through a flooded section in the main passage while another team of cavers dug the clay out of a dry high-level passage and found their way back to the same stream cave. The 'wet' and the 'dry' teams shared the excitement of the exploration that day. But still there was no way through to the huge stream passages of Speedwell Cavern – which formed part of the same system but had been explored earlier via a mine tunnel which had broken into them. So the cavers continued to probe every corner of Peak Cavern. Then in 1959 a new chamber was discovered and a narrow hole in its floor looked as if it could lead towards Speedwell Cavern. A young caver, Neil Moss, slid down the hole using a wire ladder for handholds. It was tight for him, and a corkscrew bend made it even more difficult. Less than thirty feet down he could go no further – but then found that he couldn't climb

out again. His friends tried to pull him up on the ladder but it too was jammed, and they tried pulling him up on a rope, but only managed to break it. They were helpless, though they tried everything they could; and Neil Moss too was helpless – two days later he died from the build-up of carbon dioxide in his little tube.

Cheddar and Wookey, the two main resurgence caves, are the best known in the Mendip Hills. The Cheddar Caves are beautifully decorated, but are quite short and mostly commercialized; Wookey Hole has a dry entrance section, now also a show-cave, and beyond there is almost entirely flooded – so that its exploration is in the hands of the cave divers. Mendip's attractions for the caver are on top of the hills where the streams sink underground, to eventually emerge at Cheddar and Wookey. Swildon's Hole is one of these sinkhole caves, and its four miles of passages provide endless sport. But more unusual is the nearby St. Cuthbert's Swallet. Its steeply inclined passages extend to a depth of 460 feet, but in quite a short horizontal distance. The middle section is an incredible maze – one part is named the Rabbit Warren – of chambers, tunnels and grottoes. It seems that about half the rock has been dissolved away by bygone streams continually changing their courses and cutting ever deeper; and since then this vast piece of Gruyère cheese has been liberally decorated with stalactites and stalagmites.

The cavernous limestones of South Wales are not very extensive, but where they are cut by the Swansea Valley they contain two of Britain's finest caves. On the east side of the valley is Ogof Ffynnon Ddu – now the longest and deepest known cave in Britain. Three entrances give access to a magnificent three-mile-long stream passage, and an incredibly complex high-level maze of caves extending for another twenty miles. West of the Valley, the Dan-yr-Ogof cave system is entered where a sizeable stream emerges from a gaping cave mouth. Unfortunately most of the active stream route in the cave is permanently flooded, so that, excepting a fine series of lakes and cascades not far in, the cavers are restricted to the old high-level passages. For years the limit of exploration was a nasty constriction at the end of the 'Long Crawl'. But in 1966 a young girl caver, Eileen Davies, contorted herself through this and found the passage rapidly increased in size. The subsequent explorations were fabulously successful. Spacious and decorated passages led further and further into the mountain. Many of them had dry sandy floors which made progress delightfully easy. A hundred-yard swim through the 'Green Canal' and a pair of short climbs were the only real difficulties in discovering more than four miles of new cave.

Beneath the seemingly endless peat bogs of western Ireland, the huge areas of limestone contain some magnificent caves. In

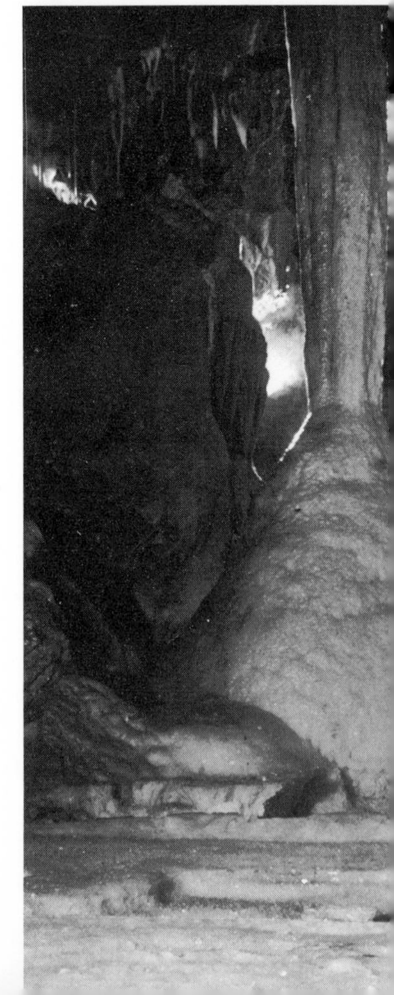

Slowly moving water and bare sculptured rock are all there is to be seen in the lowest reaches of Giant's Hole in the English Peak District (left), but further south, in the Mendip Hills, the chamber in Shatter Cave (below) is just a sparkling white mass of calcite deposits.

Ireland's Marble Arch Cave (overleaf right) is best known for its roomy stream passages, while the Sonora Caverns in Texas have the unique calcite 'butterfly' formation (overleaf left) as their trademark

the north, the Fermanagh region has everything from deep shafts to massive chamber caves and the great river passages of the Marble Arch caves. On the other hand, the caves in the second important region, County Clare, are nearly all of the one type – winding stream canyons with very gentle gradients. They are large enough for easy walking and are mostly clean washed with small, flood-prone, streams in them. The Pollnagollum system runs to more then seven miles in total length; spectacles of the area include the massive stalactite of Poll-an-ionain cave, and the Doolin cave system which contains a dry cave passage running clean underneath the Aille River which is on the surface no more than a few yards above the cave.

Until the last few years, whenever a caver has thought of American caves, the first to come to mind are the great systems of the western Appalachians, particularly in Kentucky. The geological structure and the erosional history of this area have combined to permit development of very long but quite shallow cave systems. Few are more than a few hundred feet or so deep, yet the Flint-Mammoth cave system is now over 140 miles long, and at least nine others contain more than twelve miles of passage. Both large dry tunnels and active streamways abound, and with the population spread so thinly through the area, caves are, relatively to Britain, discovered very easily. The American West has relatively fewer caves than the area east of the Mississippi. The Ozark Hills contain huge areas of limestone and deep underground drainage, together with some fine caves, while further north the Black Hills of Dakota include the huge networks of Wind and Jewel Caves. The latter gains its name from the profusion of sparkling calcite crystals which line its walls – apparently deposited in a phase when the cave was filled with carbonate-saturated water.

Probably one of the best known caves in the world lies in the Guadalupe Mountains at the southern end of the Rockies. Although originally known by native Indians, its entrance was only rediscovered towards the end of the nineteenth century, when a huge cloud rising from the opening gave the first impression of a volcanic eruption. The cloud, however, proved to be bats – millions of them on their daily exit flight to their nocturnal feeding grounds. So then it was the guano miners who became interested in Carlsbad, and they extracted huge tonnages from the entrance chambers. One of the miners, Jim White, was curious enough to wonder what lay further on in the huge passage, which sloped down into the darkness. In his spare time he took lanterns, and probed ever deeper on his solo explorations. Repeatedly he came back to tell his friends of fabulous chambers and endless caverns – but few believed him. Eventually he persuaded some to go with him, and then the stories spread across the country. For Jim White's discovery was magnificent, by anyone's standards. A few years later, in 1923, the Carlsbad Cave National Monument was created, and the cave has now been developed for tourists. It takes a couple of hours to walk round just the perimeter path in the Big Room – the largest of the superbly decorated chambers that Jim White first explored on his own.

Canada has long been regarded as an essentially non-caving country – but how wrong this has proved to be. During the last decade, systematic investigation, mainly by ex-patriate British cavers, has revealed the fabulous potential of the Canadian Rockies. In Jasper Park, the Maligne River sinks in the bed of a lake and reappears from massive springs ten miles away and 1350 feet lower with a flow-through time of only three days. This serves to indicate the scale of some of the spectacular underground drainage in the Rockies; in this case however the local cavers have been trying for years, but have still failed, to gain entry to the 'Submaligne Cave'. Further north, Arctomys Pot lies on the slopes of Mount Robson, over fifteen miles from the nearest road; consequently one of the most recently explored, it is now known down to a sump at a depth of 1720 feet. Down near the U.S.A. border, the great cave tunnels of Yorkshire Pot and Gargantua cut through the high limestone ridge of the Continental Divide. But the real centrepiece of the Canadian caves is Castleguard.

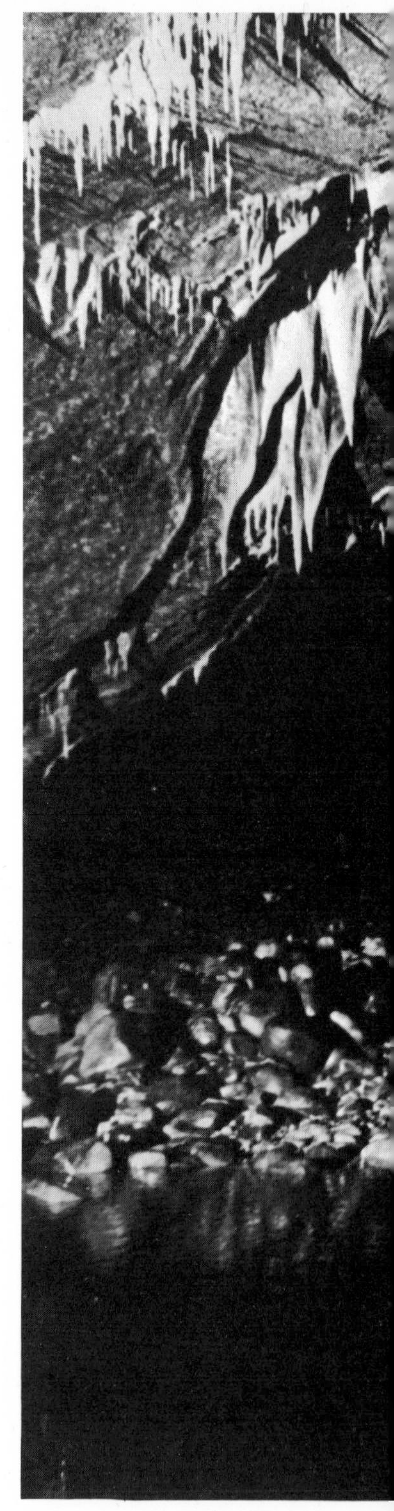

Sloping mud-covered ledges are the only footholds for the explorer carrying his pack of equipment through the deep fissure in Canada's Castleguard Cave

The Columbia Icefield is the largest sheet of ice in Canada and sits on a high plateau surrounded by limestone mountains – one of them named Mount Castleguard. Down at its foot, a river emerges from a spring and flows away to join the glacial meltwaters. Not far above it, an open cave entrance lies half obscured in the pine woods. When the cavers first visited this in 1967 they had little notion of what was to follow. The single passage climbs up into the mountain, its steady rise only being interrupted by a few downward shafts. After the first short pitch there is a 250 yard stretch of passage only passable by sliding flat out over the ice floor. (In summer this is a series of deep lakes, but the danger of summer meltwater flooding the cave restricts all the exploration to the winter.) Further in there is a mile of cave whose walls taper down to an impenetrable slot – there is no real floor; the only way along this is an arduous series of climbs along narrow ledges and traverses. The main explorations stopped in 1967, over five miles in and nearly a thousand feet above the entrance. There a small upward shaft was unclimbable with the equipment they had with them. But at that point, there was still 500 feet of rock above them, and on top of that was 500 feet of ice – for the cave runs right beneath the centre of the Columbia Icefield. An open passage was visible at the top of the climb, and they began to wonder if the cave could perhaps lead out to the bottom of a great crevasse in the Icefield. So in October 1970, Mike Boon, the English caver then living in Canada, revisited Castleguard on a mammoth solo exploration. He walked in over the snowfields alone, and then disappeared into the cave for seven days. Moving his camping gear with him, he eventually reached the shaft and climbed it using pitons – a magnificent feat – and, once up, explored an easy passage to an ice blockage. This wall of sparkling glass-clear ice appears to have been squeezed into the cave passage from the base of the thousand feet thick Columbia Icefield – a remarkable end to a remarkable cave.

South of the Rio Grande there is a whole world of caves. Central America and many of the Caribbean islands have vast areas of limestones, and fabulous caves continue to be discovered there. Cuba, Jamaica and Puerto Rico all contain major caves; Cuba's Santo Tomas cavern is one of the world's longest with nearly twenty miles of passage. The Cockpit Country of Jamaica is best known for its weird surface topography – the close packed conical limestone hills, each about three hundred feet high, resemble a gigantic egg box and make communications in the region notoriously difficult. The drainage is all underground, but mainly by flooded passages leading to deep spring pools. At the top ends of the systems some of the sinking rivers have been followed down spectacular clean-washed caves to where they sump, and some really large abandoned caves are known just above some of the springs – Windsor Great Cave is the best known.

Ten years of exploration, mainly by American cavers, have shown that Mexico is one of the world's great karst regions, and that the Sierra Madre Oriental provides some magnificent caving. The largest cave passages are mainly old and dry, but now decorated with great numbers of really massive stalagmites and stalactites; the Gruta de Palmito is only one of many examples. A complete contrast is provided by the active caves. Large streams and rivers, clean-washed canyon passages and successions of waterfalls alternating with deep pools and lakes seem to be characteristic of the Mexican caves. More than a dozen descend to depths greater than a thousand feet, and a few even

provide complete through-trips. The cave of El Chorreadero carries a sizeable stream for most of its length, down a series of cascades and through numerous lakes. Nearly two miles long and 1100 feet deep the trip in at the top and out at the bottom provides one of the world's wettest and most sporting underground journeys. It is difficult indeed to select the real *pièce de résistance* of the Mexican caves. Perhaps it is the vast open shafts in the north; the Sotano de las Golondrinas has a surface opening a hundred yards across, and its walls overhang all the way round to a depth of 1230 feet, while El Sotano – 'The Hole' – has walls which are merely vertical, to a depth of 1345 feet. On the other hand maybe the Huautla area at the southern end of the range, is the most remarkable. Adjacent to this tiny village, three deep blind valleys carry streams down to open cave systems. Between 1965 and 1968 exploration of a whole series of caves went ahead with spectacular ease. The Sotano de San Agustin ended up as the deepest after being followed down a series of huge galleries and deep waterfall shafts to a depth of 2006 feet.

A massive stalagmite boss rears towards the ceiling of a chamber in the Friouata Cavern in Morocco

Further south the caves of Guatemala are proving to contain even larger rivers than those in Mexico – and the country is still wide open for exploration. But beyond Guatemala, the flooded pots of the Yucatán and a few caves in Honduras, the underground world of South America is still relatively unknown. Expeditions of French, English and Polish cavers have explored caves in Brazil, Peru and Venezuela. Although some large river caves have been found in Brazil, none descends to any great depth. In contrast the extensive limestones of the Peruvian Andes offer considerable potential for deep caves, besides containing most of the world's highest altitude caves. In 1972 a British team explored the Sima de Milpo to a depth of 1335 feet, down a small steeply graded passage with many small pitches but no large shafts. The whole area where they worked, some 13,000 feet up in the Andes behind Lima, has great expanses of limestone with good prospects of holding even deeper caves.

Most exciting of the South American countries, with respect to caves, now appears to be Venezuela. A British expedition there in 1973 found some massive cave passages and numerous spectacular shafts in the Mérida mountains in the north-west of the country – but the deepest only descends just a thousand feet. The caves they found were already the residences of thousands of bats and guacharo birds; the latter are particularly common in Venezuela and also live in the well-known caves named after them in the north-east of the country. Some huge river caves have also been reported in the remote areas next to the Columbian border, but even more fascinating are the jungle-covered limestone plateaux in the southern part of the country, in the Amazon basin. There the ground is broken by a number of immense circular shafts; up to 1000 feet across and 800 feet deep, they strongly resemble some of the great Mexican shafts, and have the added possibility of leading down into substantial river caves.

While South America is just beginning to yield the secrets of its caves, it is tempting to wonder what remains to be found in the remoter parts of that equally vast continent – Africa. Most of Africa appears to be rather lacking in limestones, but around the edges of the continent in particular there is enough of the rock to provide some spectacular cave regions. Exploration to date has proved most fruitful in the Atlas Mountains of both Morocco and Algeria. The latter country has a number of important caves; any list is dominated by the Anou Boussouil where a series of great dry shafts descend steeply to just over 1650 feet in depth. In Morocco most of the major caves are grouped in the Taza region; the best known are those of Friouata and Chikker, between them carrying the drainage out of the large polje. In the same area the Kef Toghobeit has long been known as a rather unpleasant, loose boulder strewn series of passages leading to a sump nearly 1300 feet down; but in 1971 and 1972 French teams of cavers explored another series of passages which by-passed the sump and led down well beyond 1800 feet in depth.

At the other end of the continent, South Africa has a number of caves. Most are in the Rand region around Johannesburg, and although none is particularly long or deep they are extremely well decorated, with great amounts of calcite formations. Of a similar type are the now commercialized Cango Caves, nearer to Cape Town, where the massive decorations provide a fine tourist spectacle. Between the extremes of the Atlas and the Rand rather less is known about the caves of Africa. Rhodesia has its famous Sinoia Cave with its deep lake which has been the scene of some spectacular diving, and Tanzania too contains a number of caves. More impressive dimensions are met in Madagascar, where there are a number of fine river caves exceeding three miles in length. Kenya has a host of lava caves, but few in limestone, and Chad keeps providing rumours of spectacular cave country, though at present it is Ethiopia which can claim the finest of caves in central Africa. Caves occur in a number of regions in Ethiopia, but most are rather short and shallow; the exception is Sof Omar – a cave ten miles long located 150 miles south-east of Addis Ababa. Most of the passage length is taken up by a complex maze and network of passages on two main levels – the limestones are flat lying and there is little depth to the system. But the cave has been formed by the River Web, a main artery of the regional drainage; the entire river still flows through nearly a mile of the cave, in a passage averaging thirty feet high and fifty feet wide. Most of the floor of this impressive gallery is covered by the River

The boulder strewn tunnel of Koonalda Cave stretches beneath the limestone desert of the Nullarbor Plain in southern Australia

Web; following the water through Sof Omar provides a magnificent trip, dominated by soaring, clean rock walls and massively sculptured chambers and ceilings. The cave is almost devoid of calcite formations, but does contain large populations of bats and insects; it is convenient for the visitor that the crocodiles, which live in the river just below the entrance, appear to dislike venturing into the cave.

Australia is another country where cave exploration is hindered by the great distances between towns, cities, and even roads. The relatively well populated south-eastern part of the continent does contain a number of caves, with the commercialized Jenolan Caves perhaps the best known. Most spectacular of the mainland karst regions is the Nullarbor Plain. Flat, featureless and nearly barren, mostly less than 500 feet above sea level, the Nullarbor seems an unlikely place to find caves; yet it is formed mainly of limestone. Scattered across its vast area are great collapse holes, many of which lead down to huge horizontal cave passages. Some are at the water-table and contain long lakes curving out of sight beneath wide vaulted ceilings. Mullamullang Cave is over six miles long and the main tunnel averages 120 feet wide and nearly half as high. It slopes gently down to a depth of over 400 feet and its smooth profile is broken by great chambers containing boulder piles 100 feet or more in height.

The island of Tasmania contains Australia's most important caves. The interior of the island is mainly covered by dense rain forest, which makes access and exploration very difficult, but once an entrance is located it commonly leads to a fine cave system. Exploration is still active in Tasmania, and the most spectacular region appears to be around Junee where a large resurgence is fed by a number of sinking streams. One of these goes down the Khazad-dum pothole: an exciting series of stream

From the entrance of the Yangshuo Cave in southern China there is a fine view of the limestone towers whose walls rise vertically above the flat river plains

passages and waterfalls leads to a sump at a depth of just over 1000 feet, unfortunately before meeting the main underground river of the region. However, elsewhere on the island, through trips from sinkhole to resurgence cave are possible; one of the finest is down the great shafts of Mini Martin via a ten-mile long network of caves to the massive passages of Exit Cave, which return the caver to daylight 700 feet below the upper entrance.

The best known cave in New Zealand is Waitomo, with its spectacular glow-worm colonies hanging from the chamber roofs. It lies in a small region of shallow but interesting caves south of Hamilton in North Island. To the sporting caver, however, it is completely eclipsed by the deep Alpine-type caves of the Takaka-Mount Arthur region in the extreme north of South Island. There, Harwood Hole is the deepest; an entrance shaft nearly 600 feet deep is the way into a fine cave system leading right through to a lower entrance 1225 feet below. Five more caves in the area reach depths greater than 800 feet, and they are characterized by deep shafts and steeply descending stream passages.

North of Australia lie the great chains of islands where the Western Pacific gets tangled up with southeast Asia. Japan is not rich in limestone and its one major karst area lies at the south-western end of Monshu island. A single deep cave system reaches down to 1300 feet, while a national park has the spectacular decorated cave of Akiyoshi as the centrepiece of a fine karst landscape. The relative lack of limestone on Japan is more than compensated for by the extensive karst of Indonesia. Of the cavernous limestone regions of Java, Timor, Celebes and Borneo, it is those in the latter which have the finest known caves. Along the north-west coast of Borneo, in Sarawak, the limestones nearly all stand up as hills in low-lying regions of non-limestone rocks. This results in many large rivers flowing right through the hills in really large cave systems with relatively low gradients. Niah Great Cave is probably the most famous, for its two miles of passages average 200 feet high and wide; streamless and boulder-strewn, its floor deposits of guano have been mined for years. Consequently it is not as spectacular as the many miles of clean-washed river passages in other caves in the same region.

Of all the islands, New Guinea is now the one that appears most attractive to cave explorers. Roadless and extremely mountainous, covered in dense jungle, washed by one of the world's highest annual rainfalls and in places still only inhabited by cannibal tribes, it provides every kind of hindrance to exploration. But it contains huge areas of limestone, and recently some of the more adventurous Australian cavers have started to probe its secrets. Great plateaux scored with sinkholes have been found; there is ample relief in the limestone to conceal the world's deepest caves. So far three caves have been explored to more than 1000 feet deep, and Bibima Cave reaches a depth of 1600 feet. But these are all dry abandoned cave systems; the active river caves pose immense problems, because of the quantities of water they engulf. In a remote forested valley lies the Atea River Sink, claimed as one of the most awe-inspiring in the world. A great foaming white river thunders over an eighty-foot drop into the cave mouth. Vertical walls rise straight out of the torrent, and, short of laboriously bolting a traverse along them, one wonders if the cave can ever be explored.

The East Indies' profusion of karst and caves extend onto the mainland of south-east Asia. Malaya, Burma, Thailand, Laos and Vietnam all contain spectacular caves. Malaya's finest are probably the vast dry temple caves of Batu and Ipoh, but the other countries are more characterized by massive river caves. It is reputed that in Burma the underground drainage patterns are determined by throwing teak logs into the sinkholes and looking to see which resurgence cave they appear from! Further east lies the greatest karst region in the world – in South China.

The massive entrance of the Tunnangho Cave (right) carries a river deep into the limestone of the Kweitchow Plateau in the karst of southern China.

Stalactites, stalagmites and flowstone characterize the water passages of the Krizna Jama cave (opposite above) in Northern Jugoslavia. In contrast, the bare walls and sand floor of the imposing tunnel in Fitton Cave (opposite below) is typical of many caves in the Eastern USA

The dramatic tower karst of Kwangsi is deservedly famous, but the caves developed in these limestone pinnacles cannot be of any great extent. Instead a caver would turn to look in the Kweichow Plateau limestones. There is a sad lack of information on caves coming from this part of the world, but the few visitors to the region have returned with glowing reports of immense rivers disappearing into mammoth cave entrances. It appears that the golden age of Chinese cave exploration is yet to come.

Greatest of all the world's mountain chains, the Himalaya does not appear to contain the same proportion of cavernous limestones as, for example, the Alps. The summit block of Mount Everest, as well as the main bulk of both Dhaulagiri and Annapurna, consist of limestone, but both geological and climatic factors mean that man-sized cave systems are not to be found there. Further west, Nanga Parbat claims the distinction of the world's highest known cave; Rakhiot Cave lies at an elevation of 21,600 feet, but consists only of a single gallery a few hundred feet long. There are of course vast areas of the Himalaya which have never been visited by cavers, and parts of Tibet and Pakistani Kashmir would seem to offer some promise. At present the only known large caves in the Himalaya are at low altitudes, in the foothills. Both Assam and Nepal contain some short but spectacular river caves.

Mexico's warm climate has aided the rich growth of massive stalagmites in the Cathedral Room of the Palmito Cave (above). The Krizna Jama cave (far right), in Jugoslavia, contains a whole chain of lakes which have to be crossed with portable boats. The main obstacles in the English Pennine caves are the shafts; a visitor to Lost Johns Cave (right) waits for his friends to rejoin him

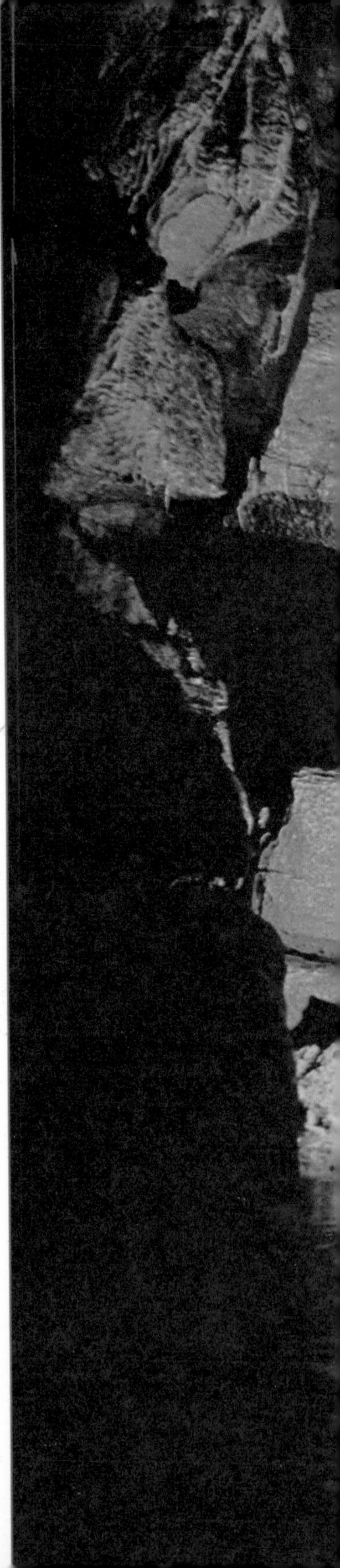

Under a series of different names, the mountain range which includes the Himalaya continues westwards right across Asia. In Pakistan and Baluchistan the peaks are lower and the limestones are unexplored, but in Iran some recent explorations have had spectacular results in the Zagros Mountains. Two British expeditions, in 1971 and 1972, explored the limestone blocks around the town of Kermanshah. Sinkholes near the summits have their only possible outlets more than 5000 feet below, where large springs issue from the limestones to provide rivers flowing across the adjacent plains. Unfortunately the one cave explored to any significant depth, Ghar Parau, consisted of just a single, narrow steeply descending, arduous passage which led to an impassable sump only 2460 feet down. However this part of the Zagros contains some spectacular high-level doline fields, and once the problems of exploring these in the vicious desert heat have been overcome, there is every chance of finding caves to take the world depth record to Iran.

Over the border into Iraq there are at present little more than rumours of karst and caves, but further west again, Lebanon has been aptly described as the 'Land of Limestone'. Most famous of its many caves is Jeita – now turned into a show-cave, and not far from Beirut. It is a resurgence cave, carrying a really large river, and its passages are magnificently decorated with masses of both large and small calcite formations. Jeita in only one of a number of large resurgence caves which feed the water of the mountains out towards the sea. And the source of all this water is to be found on the high limestone plateaux which are riddled with hundreds of shafts and potholes. Five of these reach depths of more than 500 feet, and the deepest, Faouar Dara, has a sump at a depth of 2040 feet, reached by a magnificent twisting canyon passage. Clean-washed walls soar upwards, and the cave roof is lost in darkness for the whole length between the entrance and sump; a whole series of shafts, almost dry in summer, alternating with deep plunge pools and lakes, make the descent of Faouar Dara a classic.

The Taurus Mountains of Turkey also contain great areas of limestone, much of which is carved into spectacular karst. Underground drainage in the area is on a grand scale, with large closed drainage basins feeding to massive springs, in some cases dozens of miles away. Years of exploration, mainly by French cavers, have not really produced cave discoveries on a comparable scale. A few short and spectacular river caves with really large passages are known, but the deepest cave, Dudencik, only just passes the 1000 foot mark. One of the longest caves has been explored upwards from the resurgence. This is Pinargozu, which carries a large stream, but not many yards inside is almost sumped. The narrow airspace left open develops a hurricane

Famous for its decorations is the Hall of the Thirteen, 1500 feet down in France's Berger Cave

force wind, as all the cold, heavy, air in the cave flows downwards into the warmer outside. The water in the pool at this constriction is blown into waves which make passing it unusually difficult. Further inside, a succession of difficult climbs up the cold cascades give access to miles of fine passage, and exploration has halted where the water is met sheeting down high unclimbable shafts.

The immense land area of Russia does of course include extensive areas of limestone, but cave exploration in the country only appears to have really progressed in very recent years. Most important of the cave regions in Siberia are around Irkutsk, and one of the finest caves there is the Kubinsk Cavern – over 800 feet deep and a mile and a half long. A series of shafts lead down to passages which are beautifully decorated with stalactites, stalagmites and cave pearls, and contain a lake in one chamber. The Urals contain numerous caves too, including the Sumgan Cave, where a 200 foot shaft leads to over two miles of passage. Even more exciting is the Alaya region further to the south. There, many caves exceed two miles in length and 500 feet in depth, and some are noted not only for the stalagmite decorations but also for their spectacular linings of calcite crystals. Russia's cave masterpieces however are all in the western part of the country. The Caucasus and the Crimea contain all the nation's deepest caves, including one recently explored to a depth of over 2400 feet. In complete contrast, all the longest caves are found in the Ciscarpathian karst, even further west, formed not in limestone but in gypsum. The cave of Peschtschera Optimititscheskaja consists of a maze of passages totalling sixty-five miles in length – and is the world's third longest cave system.

At the present time it is the Alpine mountain chains of western Europe which contain the greatest number of the world's major caves. Cavers tend to judge a system more on its depth than its length, and Europe contains forty-four of the world's fifty deepest caves. This is at least partly due to the geological influence on cave patterns. While the large areas of flat-lying limestones in places like the U.S.A. provide the ideal environment for very long cave systems, the very deep caves are found in the young mountain chains. There the limestones are folded up to great heights and of course there is the height difference between summits and valleys to accommodate the cave depths. Not only are the European Alps rich in limestone but they are also of course the most accessible and best known mountains. So until exploration reaches a comparable level in such regions as the Zagros Mountains of Iran or the Star Mountains of New Guinea, Europe will always be the mecca of world caving.

Nearly every country in Europe has its own major caves, but it

Deep in the Zagros Mountains of Iran, the second pitch of the Ghar Parau pothole takes the caver down into a chamber containing a massive stalagmite boss

is the northern ones, away from the Alps, which are least well endowed. Britain has its few small areas of intensely explored cave-bearing limestones, and is much better off than Scandinavia. There, only Norway has any sizeable caves, and these are all in the thin marble bands folded into the mountains of Nordland. Stalactites are almost completely lacking due to the cold climate, but this is more than compensated for by some of the spectacular stream passages cut into the polished hard white marble. Visiting English cavers are continually finding new caves in Nordland; at present the finest is the 1885 foot deep Ragge-javre-raige pothole which provides a through-trip from the high fells almost to the very edge of a fiord.

The low countries of Benelux and northern Germany contain very few caves, and it is a remarkable anomaly that in the centre of this region lies one of Europe's finest show-caves. At Han in the Belgian Ardennes, the River Lesse passes right through the centre of a relatively small limestone hill, and over a long period of time it has cut a magnificent series of caves. The old dry passages are now spectacularly decorated, and some of the presently active passages are flooded, but at the downstream end of the system the whole river flows out through a magnificent gallery forty feet wide and twenty feet high; the walls drop straight into deep water, so the only way down the passage is by boat and straight out into the open air. In northern Germany there are only a few caves in gypsum near Hanover, and the southern part of the country is little better off. The Schwabian Alb contains a handful of active and inactive caves, while the only deep systems are in that little section of the limestone alps 'borrowed' from Austria, where the frontier loops round the Berchtesgaden valley in southern Bavaria.

The Eastern European countries – Poland, Czechoslovakia, Hungary, Rumania and Bulgaria – all have fine karst regions with numerous caves, many reaching depths greater than 600 feet. Rumania has a number of systems with large streams flowing in large gently graded passages, and two reach depths of more than 1000 feet. The deepest of all the East European caves is located in the high Tatras Mountains of southern Poland; the Jaskinia Sniezna consists of a series of cold wet shafts and narrow meandering canyons which descend steeply to 2466 feet below its upper entrance. The Tatras karst continues over the border into Czechoslovakia where the Demanova group of caves are the most magnificently decorated of that country's many significant caves. Demanova includes a whole series of separate caves in the wall of a valley draining south off the Tatras mountains. A river occupies the lowest passages, higher up are some beautiful stalactite caves, and above them is the dramatic Demanova Ice Cave. The Moravian karst in the centre of the country includes

In the Belgian Ardennes, the River Lesse flows out of the gaping entrance of the Han Cave (left). The Père Nöel Cave was formed when the Lesse took a different course through the same hill; now it stands dry and abandoned, but beautifully decorated by a mass of calcite deposits (above)

A cross-section through the Epos Chasm in Greece shows the three deep shafts which drop to the final lake

mainly smaller caves, and the outstanding spectacle is the Punkevni Cave. Two passages – a dry series of decorated chambers and an active river cave – lead from a low-level entrance into the hillside and open into the side of the Macocha Chasm – a 1000-foot-long and 400-foot-wide shaft open to daylight 400 feet above its rock strewn floor. However, Czechoslovakia has to share its longest cave with Hungary – the great Baradla-Aggtalek cave system extends for nearly fifteen miles of passages between a total of seven entrances on both sides of the international frontier.

The many large clear springs, which were revered in ancient times and helped to make Greece a pleasantly inhabitable region, nearly all emerge from cavernous limestone. Much of Greece, both mainland and islands, is made up of limestone and numerous caves have been explored. But perhaps the most remarkable of all is a completely flooded cave where the water flows uphill! This is in the island of Kefallinia, where at Argostolion, on the west coast, the sea flows into a series of sinkholes on the foreshore just three feet below mean sea level. A mill has been constructed here with the wheel turned by the stream flowing landward. This water has been proved, by dye-tracing, to emerge from springs on the opposite coast of the island, twenty-four miles away and three feet above mean sea level! The mechanism behind this apparent mystery is that part of the way along its journey through the flooded cave beneath the island, the sea water is mixed with some freshwater draining down from the limestone mountains 5000 feet above; the springwater is in fact noticeably brackish. The shape of this cave must be visualized as a massive U-tube; the upstream column of heavy sea water is then balanced by a higher, downstream column of lighter brackish water, giving the overall, uphill flow.

Most exciting of the cave explorations on mainland Greece have been those in the Pindus Mountains. British cavers were attracted there in the late nineteen-sixties by the stories of Provetina, the great single shaft which was eventually found to be 1285 feet deep. The last expedition to Provetina also found the entrance and explored the first section of the Epos Chasm no more than a mile away over the Astraka Mountain. A return expedition to Epos in 1969 found that it was an almost totally vertical system. The great gash on the hillside led directly to a succession of three vertical shafts, 445, 550 and 305 feet deep. It was a spectacular descent, but rather a disappointment when the cavers descended the last shaft and found the smooth walls, and the ladder, dropping straight down into the black waters of a deep lake; there was no way on past the 1450 feet depth level, even though the water only re-emerged on the surface yet another 2000 feet lower down.

Over the border into Albania the limestones of the Pindus continue, but little is known of caves in the mountainous karst of this rather inaccessible country. Further north still the picture is different, for Jugoslavia is not only the home of karst but is famous around the world for its caves. The coastal mountains along the length of the country, from Dalmatia to Istria and Slovenia are almost entirely limestone. Though the southern half contains really superb karst, with massive poljes draining underground (some to large submarine springs), there are very few known caves of any significant size. Slovenia, which includes the Kras – the classical karst region – contains thousands of caves, most dramatic of which are the great river caves which carry the drainage underground between the poljes. The River Reka disappears into the enormous passages of the Skocjan cave and is then seen briefly in three places, all at the bottom of deep shaft caves, on its thirty-mile journey under the frontier to the great Timavo springs near the Italian coast.

The most famous Slovenian caves are those of the Postojna group, where grand river caves are found beneath great series of old abandoned passages, now beautifully decorated with stalagmites and stalactites. The main Postojna cave even became involved in the Second World War. The German occupation of the area led to the cave being used as a munitions store by the occupying military forces. With a well guarded entrance, the military did not however appreciate the dangers from within the cave; a group of caving patriots entered the cave by a second entrance deep in the forest, and made their way through the miles of passages known only to them. It was then an easy and undisturbed task to lay charges to destroy the stored munitions, before retreating along the way they had come in. To this day, visitors to Postojna can see the smoke-blackened walls and stalagmites, as they ride the miniature railway through these entrance passages into the decorated chambers in the heart of the system.

The geology of the Alps is such that right in the core of the range is an area relatively poor in limestones, compared to the surrounding mountains. So Switzerland has never become a caving region to match France, Italy and Austria – even though right in its centre, near Lucerne, lies one of the world's longest and deepest caves. Hölloch consists of a great sloping maze of roomy tunnels and vast chambers. It has one entrance, down near its resurgence, but from there climbs steadily into the mountain, fed by trickles of water from the bare Muota plateau above, but nowhere containing a large cave river. Altogether seventy-five miles of cave passages are known extending over a vertical range of 2650 feet. Hölloch tends to overshadow Switzerland's other caves, even though there are at least eight others

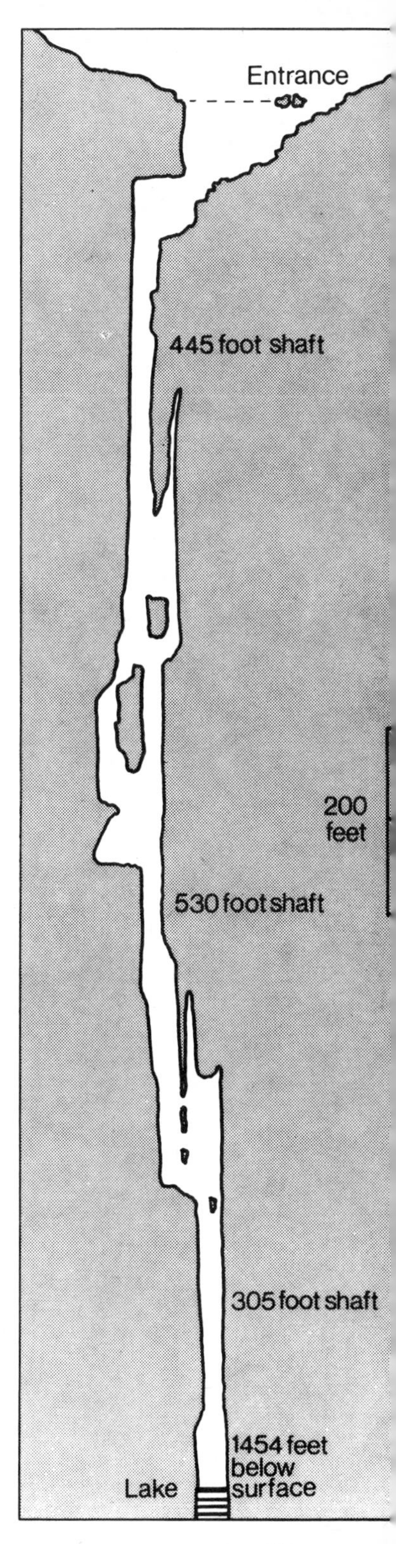

Spacious galleries strewn with limestone blocks are typical of the upper series in Italy's great Monte Cucco Cave

which exceed 1000 feet in depth, and the limestones around such well-known places as Interlaken still have much to offer the dedicated cave explorer.

Massive river caves, spectacular potholes and decorated show-caves are spread right across the mountains of Austria; but the greatest concentration is in the Salzburg Alps. This impressive area of bare Alpine karst is famous not only for its deep caves, such as the Gruberhornhöhle whose shafts plunge downwards for a total of 2330 feet, but also for its ice caves. There are many of the latter, but much the best known are those of Eisriesenwelt and Dachstein: both are extensive systems of large dry boulder-strewn caves, and both have their entrance sections adorned with ice (and now opened up as show-caves). The ice is on a grand scale – towering stalagmites and enormous cascades of solid white ice decorate the spacious tunnels, some of which have veritable glaciers along their floors.

Italy is second only to France as a country of deep caves, and half a century ago it contained nearly all the world's deepest caves known at the time. It was Italian cavers who first passed the 1000 foot depth limit when in 1841 they explored the Trebiciano chasm in that part of the classical karst which overlaps from

In Austria's incredible Eisriesenwelt Cave the main tunnel is all but filled with ice, leaving a route around its edge past this great frozen wall

Jugoslavia. 1927 saw the first deep exploration into the great Spluga de la Preta high in the Verona Alps. Plunging shafts of 430, then 350 then 280 feet gave the cave the depth record at the time, and it was a magnificent feat to master these shafts with the crude lighting and heavy rope-ladders then available. Since then the Italian cavers had repeated successes all over their country, across the Alps, down the Apennines and even in the far southern corner of Calabria.

One of the most remarkable karsts in Italy is on the Marguareis Massif, overlapping the French border just north of Nice. The deepest cave there is named the Piaggia Bella, which translates as 'beautiful place'. Unfortunately the full potential of the cave has yet to be realized, and even though it now has three entrances to its fine streamways and galleries, it cannot be explored beyond a sump 2260 feet down, still well above its resurgence. The Julian Alps extend as a limestone range in both Italy and Jugoslavia, and now contain the deepest known caves in both countries. Italy's number one is the Abyss of Michele Gortani. Exploration between 1965 and 1970 revealed that its miles of galleries connect countless shafts, which eventually lead to just beyond the 3000 foot depth mark.

Both the Alpine mountain chains and the limestones which form part of them, continue westwards as far as the Atlantic Ocean – along the line of the Pyrénées and then the Cantabrians. Spain is therefore rich in caves, and indeed not only in these areas, for limestone near Madrid and Málaga both contain important caves and in addition there are the Balearic Islands. The main mountains of Majorca are formed of limestone, and recent explorations have revealed a number of deep potholes in the area, though as yet they cannot match the dramatic appeal of the commercialized Drach cave, with its underground lake and enormous numbers of stalagmites. It is difficult to select any outstanding caves in the Cantabrians – there are so many of all possible types. Pride of place perhaps should belong to the painted caves of Altamira, though the caver will prefer the active stream caves higher in the mountains. Valporquero is a cave which is a pleasure to explore; a mile long and 450 feet deep, its clean-washed stream passage can be followed from sink to resurgence. Further east the cave system of Ojo Guarena contains over thirty miles of passages, and the Torca de Carlista has a 500 foot deep pitch leading into a chamber often claimed as the largest in the world; 1300 by 800 feet, and 300 feet high, it is less broken up than the famed Big Room of the Carlsbad Caverns. One of the deepest of Spain's caves is the Gouffre Juhué, further east still where the Cantabrians merge into the Pyrénées. Just inside its entrance is a 990 foot deep shaft followed immediately by more shafts which eventually break out into a two-mile long dry horizontal gallery. A hole in the floor of this leads to the deepest point 2540 feet below the entrance. The Pyrénées themselves are a great limestone range, but chance has it that most of the big caves are on the French side of the watershed frontier.

Any account of the world's caves must culminate with France. Not only can the French claim five out of the seven deepest caves known in the world today, but it has over sixty others which reach depths of more than a thousand feet, besides every possible variety of pothole, cave system, river cave, decorated cave, and painted cave. The Jura, the Causses, the Ardêche, the plateaux of the south-east – all are magnificent cave regions, yet they are overshadowed by both the Pyrénées, and the French Alps (culminating in the Dauphiné and the Vercours). France is the land of the 'super-caves', and the story of French caving is one of protracted and arduous, but fabulous explorations thousands of feet below the surface.

Fifteen years of successful exploration in the Arbas mountain in the Pyrénées have at last revealed the extent of the great Felix Trombe cave system. Ten different potholes have been explored all down the flank of the mountain. Each has led to networks of stream passages and high-level abandoned galleries.

Towering stalagmites loom out of the darkness in the Hall of the Thirteen in France's Berger Cave

Continuing probes by the cavers have found more and more connections between the separate potholes, so that now the whole series forms one giant system with twenty miles of passage over a vertical range of 2880 feet.

A very different pattern of exploration developed in the Aiguilles cave system in the Dauphiné Alps. Cavers were shown the entrance by a shepherd in 1965 and a series of summer expeditions took them to a depth of just over 2000 feet by 1969. The passages were narrow and difficult; loose rock was an extra hazard, and the sump they found was impossible; but 600 feet down an inlet passage joined the main cave. So in 1972 cavers from Toulon took up the challenge and started to explore the inlet – upwards. Their new passage at first ascended slowly giving them an easy route up through a series of boulder chambers; but beyond it developed into a fine canyon passage carrying just a small stream. They continued, climbing a whole succession of shafts, the highest being well over 100 feet. A year's exploration took them much higher in their inlet than the entrance by which they always entered the cave. Eventually they reached a spot where they dug a few rocks and boulders from the end of their cave and emerged yet again in the open air. Their new entrance was much higher up and 3090 feet above the sump.

Not surprisingly the world's deepest cave is in France – but only just, for it lies in the Pyrénées and actually extends under the frontier so that one of its entrances is in Spain. The Pierre St Martin cave was first entered in 1951 by a large expedition which had to spend most of its effort winching cavers up and down the Lepineux shaft, which dropped a full 1050 feet from the surface. They returned the next year, but made little progress partly due to the untimely death of Marcel Loubens when the winch failed, and in 1953 they came to a stop in the gigantic La Verna chamber where the cave river sank into a gravel choke nearly 2400 feet down. Political arguments then hindered the cavers for some years – during which time it was decided that the Lepineux shaft lay just in Spain. Then in 1960 a hydroelectric diversion tunnel (obsolete before it was completed) gave cavers an easy route into the La Verna chamber, and a series of new discoveries soon followed. A roof passage in the great chamber led to more descending shaft systems which end in impenetrable fissures, but more important were the upstream discoveries beyond the foot of the Lepineux shaft. Nearly three miles of cave, stream passages, dry galleries and lake chambers, led the explorers way beneath a new high-level region of limestone benches and scars. And in 1966 the Tête Sauvage pothole was explored down a steep series of shafts into the main gallery of the known cave. The total depth of the system now stands at 3850 feet; and exploration still continues.

The caver's lighting can only give a shadowy glimpse of the truly magnificent grandeur of La Verna chamber in the Pierre St Martin Cave

The Pierre St Martin may be the world's deepest cave, but its three entrances make it rather too accessible; it doesn't offer cavers the attraction and challenge of a long descent into the remote depth of a cave with only one entrance. So if asked which system they would most like to visit, the majority of cavers would probably not choose the Pierre St Martin, but the Gouffre Berger.

Jo Berger, a caver from Grenoble, found the entrance of the cave that was named after him in 1953, high in the Vercours and overlooking his home-town. The local cavers with him explored a series of shafts and narrow galleries which over 800 feet down opened into the side of a vast gallery. Forty feet wide and twice as high this fabulous cave curved away into the darkness. The team's jubilant explorations were halted at a lake which occupied the whole width of the passage. They realized the cave was going to be formidably large so they resolved to return the next summer with a strong expedition. This was led by Fernand Petzl, a superb caver who not many years earlier had been one of the leaders of the nearby Trou du Glaz explorations. The 1954 expedition set off down the cave armed with plenty of ropes and ladders, a rubber dinghy for the lake, and all the equipment needed to camp underground. They climbed steadily down a vast boulder-strewn gallery, and chose for their campsite a fabulously decorated chamber. Massive stalagmites are grouped in clusters on a floor of white calcite gour dams and crystal clear lakes.

On the second day of their underground journey it was a pleasure to follow this series of gour lakes on down the cave. Further on the formations decreased in number and the overall gradient of the cave steepened down to a section where the only way on was through a series of canals. Waterfall shafts and intervening lakes made their journey slow but exciting until they entered another series of vast boulder-strewn chambers. At the bottom of these, 2950 feet below the entrance, they turned back for the year, with the sight of the cave stream plunging over a high cascade into further depths. This series of waterfall shafts, so deep down in the Berger, provided continuous difficulties for the cavers as they endeavoured to hang their ladders clear of the freezing cascades; it was two more years before an international expedition reached the sump 3680 feet down. Further discoveries in the cave, since that date, have not added to the journey to the sump, but it is still nearly three miles from the entrance.

This journey down the Berger cave is still a challenge, and involves the caver in a lot of hard work. But it is hard work and effort well spent, for the fabulous series of massive galleries, roaring streamways and decorated chambers provide a unique experience. The Berger is a classic, of course, but its fascination extends to the whole world of caves.

Further reading

A considerable range of books have been published on the subject of caves, covering all aspects from exploration through to the sciences. Consequently this list of recommended reading is very selective, and also omits any books not available in the English language.

Two volumes are notable for their broad general coverage on the subject of caves:

British Caving edited by C. H. D. Cullingford. Routledge and Kegan Paul, 1962 (Second Edition)

Radiant Darkness by Alfred Bögli and Herbert W. Franke. Harrap, 1967

The latter is a translation from the original German and contains many fine photographs of Central European caves. The former is more systematic and also acts as a useful introduction to the scientific aspects of caves.

No single book yet published is devoted entirely to the origin and formation of caves. However there are hundreds of scientific papers, in various widely available journals, which describe and theorize about the origins of caves; these are best located by referring to two other useful books:

Karst Landforms by Marjorie M. Sweeting. Macmillan, 1972

Karst by J. N. Jennings. M.I.T. Press, 1971

Both cover the whole subject of limestone scenery, and include chapters on caves and cave deposits. The former is a large and comprehensive volume with worldwide coverage, while the latter is less exhaustive, therefore rather easier to read, and draws examples mainly from Australia. There are also a number of books describing individual cave regions; one of the more recent volumes is:

Limestones and Caves of North-west England edited by A. C. Waltham. David and Charles, 1974

Life in caves has its own extensive literature, but two books are outstanding:

Biospeleology by A. Vandel. Pergamon Press, 1965

The Life of the Cave by Charles E. Mohr and Thomas L. Poulson. McGraw Hill (New York), 1966

The former is the standard reference work, while the latter has a much lighter approach, with a profusion of illustrations.

Books concerning the sporting and exploration aspects of caving provide much lighter reading than the scientific volumes. Besides handbooks on how to go caving, and guidebooks to caving regions, much of the most exciting reading is found in the various national and club journals which unfortunately have rather limited circulation. But there are also many books on caving, and the following would probably end up on most people's list of favourites:

Potholing Beneath the Northern Pennines by David Heap. Routledge and Kegan Paul, 1964

Mendip – its Swallet Caves and Rock Shelters by H. E. Balch. Simpkin, Marshall, 1948 (Second Edition)

Ten Years Under the Earth by Norbert Casteret. Dent (Aldine paperbacks), 1963

One Thousand Metres Down by Jean Cadoux. George Allen and Unwin, 1957

Subterranean Climbers by Pierre Chevalier. Faber and Faber, 1951

David Heap's book describes, in a delightful style, the joys of visiting half a dozen of the major caves in the English Pennines. Balch's book describes caving in the days of candles and tweed jackets, in the 1920's mainly. It is poetically written and gives a remarkable insight into the rigours of caving without modern equipment.

The last three books are all translations from the French, which reflects the profusion of caves and cavers in France. No caving reading list can fail to include one of Casteret's books. A forerunner of cave exploration for over thirty years, and an artistic writer, his books describe an amazing series of adventures, mainly in the caves of the Pyrénées. Jean Cadoux's book describes the exploration of the Gouffre Berger, and is therefore a classic. But possibly the finest book ever written on caving is Pierre Chevalier's. It is now out of print and so only available from libraries, but it is well worth a search. In complete contrast to Casteret's style, it is a masterpiece of underwriting, as it describes ten years of exploration in the Trou du Glaz cave system.

Index of caves

Photo acknowledgements

All the photographs used in this book are the author's own work apart from the following:

T. Aley: 97 tl and r
Ardea Photographics: 95 b, 194
D. Balazs: 28, 29, 45 t and b, 92 b, 218, 220
R. J. Bowser: 43 b
M. C. Brown: 40
Camera Press: 155 l, 183 (S. C. Bisserot), 184 (L. R. Dawson), 189 (Optik photograph), 190–1, 198–9 t (P. G. Wichman)
D. Checkley: 98, 170, 181 tl, 182 b
Bruce Coleman: 180 (J. Burton)
P. J. Collett: 75, 182 tl and tr
J. M. H. Coward: 196–7
G. Cox: 9 l, 84, 157 tl
R. D. Craig: 143 tr
J. A. Cunningham: 48 b, 52 l, 54, 106, 137, 211, 214
P. R. Deakin: 46, 51, 236
G. Egginton: 184 tl 221 t
G. Eldridge: 37 b
French Government Tourist Office: 92 t, 186
W. Gamble: 101
Gibraltar Tourist Office: 109
Dick Handley: 16
R. Harmon: 159, 210, 221 b, 222 t
J. H. D. Hooper: 174, 177 t
K. Jelley: 195
D. M. Judson: 32
Frank W. Lane: 168
M. H. Long: 8 tl, 9 t and br, 35, 82, 104, 127, 134 t, 135 t, 136, 149, 207
J. R. Middleton: 39 bl, 94 tl, 143 tl
New Zealand National Photo Library: 15, 88
J. Russon: 119 b
P. Smart: 164 b, 184 b
A. J. Sutcliffe: 25 b, 103, 165, 172–3, 178, 179, 192, 193
B. H. Twist: 233
U.S. National Park Services: 177 b
J. C. Whalley: 8 r, 42 b, 49, 115, 138, 226, 232
T. M. L. Wigley: 217
D. Wilkinson: 166–167
J. R. Wooldridge: 2 (frontispiece), 52 r, 66, 69, 223
C. Wood, 24 t
P. F. Wycherley: 198–9 bl and br

Diagrams drawn by Rodney Tubbs from originals by the author, and after survey by D. Catlin (p. 63), survey by P. Courbon (p. 72), and survey by A. N. Palmer (p. 100).

Grateful acknowledgement is given to all contributors for the use of their material

ADIRONDACK COMMUNITY COLLEGE
GB601.W34 1974
Waltham, Tony. Caves
3 3341 00016141 3

Clare
Pennines
Jura
Alps
Vercours
Causses
Cantabrians
Pyrénées
Majorca